T0301723

Program Management Leadership

Creating Successful Team Dynamics

Best Practices and Advances in Program Management Series

Series Editor
Ginger Levin

PUBLISHED TITLES

Program Management Leadership: Creating Successful Team Dynamics
Mark C. Bojeun

Successful Program Management: Complexity Theory, Communication, and Leadership
Wanda Curlee and Robert Lee Gordon

From Projects to Programs: A Project Manager's Journey
Samir Penkar

Sustainable Program Management
Gregory T. Haugan

Leading Virtual Project Teams: Adapting Leadership Theories and Communications Techniques to 21st Century Organizations
Margaret R. Lee

Applying Guiding Principles of Effective Program Delivery
Kerry R. Wills

Construction Program Management
Joseph Delaney

Implementing Program Management: Templates and Forms Aligned with the Standard for Program Management, Third Edition (2013) *and Other Best Practices*
Ginger Levin and Allen R. Green

Program Management: A Life Cycle Approach
Ginger Levin

Program Management Leadership: Creating Successful Team Dynamics
Mark C. Bojeun

The Essential Program Management Office
Gary Hamilton

Project Management in Extreme Situations: Lessons from Polar Expeditions, Miltary and Rescue Operations. and Wilderness Explorations
Monique Aubry and Pascal Lievre

Program Management for Business: Aligning and Integrating Strategy, Process, Technology, People and Measurement
Satish P Subramanian

Program Management Leadership

Creating Successful Team Dynamics

Mark C. Bojeun, Ph.D.

CRC Press is an imprint of the
Taylor & Francis Group, an **Informa** business
AN AUERBACH BOOK

CRC Press
Taylor & Francis Group
6000 Broken Sound Parkway NW, Suite 300
Boca Raton, FL 33487-2742

Printed on acid-free paper
Version Date: 20131112

International Standard Book Number-13: 978-1-4665-7709-1 (Hardback)

Library of Congress Cataloging-in-Publication Data

Bojeun, Mark C.
Program management leadership : creating successful team dynamics / Mark C. Bojeun.
pages cm -- (Best practices and advances in program management series)
Summary: "The book focuses on individuals who have come to understand the values of the tools that are provided by PMI but are still looking for the advantages and the success factors necessary to be truly great Program and Project Managers. The targeted audience is one that continues to focus on self-actualization and continuous improvement as a way of learning from historical efforts and driving each new initiative to the highest possible set of standards. "-- Provided by publisher.
Includes bibliographical references and index.
ISBN 978-1-4665-7709-1 (hardback)
1. Project management. 2. Teams in the workplace--Management. I. Title.

HD69.P75B64 2014
658.4'04--dc23 2013042121

Visit the Taylor & Francis Web site at
http://www.taylorandfrancis.com

and the CRC Press Web site at
http://www.crcpress.com

Leadership must be based on goodwill. Goodwill does not mean posturing and, least of all, pandering to the mob. It means obvious and wholehearted commitment to helping followers. We are tired of leaders we fear, tired of leaders we love, and tired of leaders who let us take liberties with them. What we need for leaders are men of the heart who are so helpful that they, in effect, do away with the need of their jobs. But leaders like that are never out of a job, never out of followers. Strange as it sounds, great leaders gain authority by giving it away.

—Admiral James B. Stockdale

Contents

SECTION II Leadership

SECTION III Leadership and Teams

SECTION IV Formal Leadership Processes

Preface

About a year ago, a colleague of mine offered me the opportunity to write a book tailored to program managers and, more specifically, on how leadership can create high-performing teams (HPTs) that regularly exceed expectations and operate as a collective, innovative, communication-driven, and conflict-positive group.

At the time, I jumped at the idea. Not only have I been working as a project and program manager for more than fifteen years, I have been teaching program/project/risk management courses both commercially and academically for ten-plus years. The idea of writing about one of my favorite subjects seemed ideal for the next challenge. However, writing this book has truly been a journey and not a dissertation. Through each chapter, case study, and example, I have finally found the opportunity to review the conscious decisions and management styles I have employed and the results of my approaches. There is no doubt that I have had the opportunity to work with some really fantastic teams that truly achieved HPT status, but I have also struggled with team development, cultures, communication issues, and conflicts.

If you had asked me a year ago about my ability to develop HPTs and lead programs to successful conclusions, I would have immediately shouted, "Yes, of course I can do that." After writing this book, I realize that so many factors go into developing a team—including each member's skills, abilities, and willingness to join a team—that to be successful, leaders not only must make conscious choices on leadership but also must be able to actively read and interact with the corporate culture and environment, and to personally invest constantly in the team. A leader will work individually and with the team as a whole to bring members together, establish trust and communication, ensure that conflict resolution is positive, and delegate authority to empower teams to achieve objectives without a micromanagement approach that involves the leader in every decision point.

Over the year's journey I undertook to write this book, it occurred to me that whereas I have evolved as a leader over the multiple programs I have managed, the team members I have had the opportunity to work with have played a crucial role in achieving the objectives we set out to deliver. In many situations, I relied more on my instinct than on professional training as to how to handle a situation or individual challenge yet with the research into this book, I now have a much more effective strategy for handling future efforts. In other words, this book was written to help program managers exceed expectations, and in the process, it has helped me to become a better leader and team participant.

Program Management Leadership: Creating Successful Team Dynamics is not a how-to for program managers or a reiteration of the Project Management Institute's standards for program management; rather, it is focused on two key points. The first is leadership, including the styles, traits, and choices that leaders make and how they work with the stakeholder and team members to set vision, objectives, and benefits management plans to ensure that programs achieve the desired objectives. The second and most important aspect of the book is the focus on teams and how to bring disparate people together who choose, for a temporary time, to set aside their personal objectives and instead work toward those of the team and toward the program manager's vision.

This book describes both the research on leadership as well as the situational factors that will require leaders to modify their style from one based on personal choice to one that can overcome the challenges faced from both individual team members as well as stakeholders and organizational-cultural factors. From that foundation, the book drives toward how to build and maintain an HPT and how to ensure that the team continues to be driven toward success through the employment of competitive spirit, cooperation, and mutual respect for each other. Although there are many tools leveraged by leaders in managing programs and projects, this book focuses specifically on the leadership approach and the impact that it has on achieving program/project success. To be completely clear, this is both a science and an art. There are a tremendous number of leadership styles and approaches available in today's world. The key to program management is the ability of a leader to recognize which approach to use when, what the impacts will be, and when to change that style to another approach that will further drive the team.

Writing this has been not only a challenging and unique opportunity, it has also been a personal journey that has taken me from my initial attempts at program management to the current efforts I manage. The three- to six-month process I envisioned evolved into a year-long effort requiring a tremendous amount of support from colleagues, friends, and family. Although the list is extremely long, I would like to thank my mentors, Dr. John Whitlock from Capella University and Dr. Roy Hinton with

George Mason University, as well as to sincerely thank my family for their support and assistance while I wrote and rewrote chapters time and time again to try to better communicate the leadership approaches and true value of high-performing teams. Thanks go to my beautiful wife, Melissa, who supported me and empowered me to continue on one of the most challenging projects of my career. While she provided emotional support she also ensured that I met every deadline and improved the book with every edit and comment she made. I honestly could not have achieved the completion of the effort without her telling me to continue, and finally—when to stop rewriting sections. Without her support and effort I would have not completed this effort and I am grateful to her for all she has done. And to my children, Austin, Lily and James, who were willing to give up family time so that I could focus on the demands of writing while still maintaining a full-time career. Finally, to my daughter who I miss so very much, Ariel, you have inspired me with your writing and career. I have learned so much writing this book that even my fifteen-year-old son, James, taught me that paybacks are fair game. Each night I dutifully asked him if he had completed his homework only to hear, "Don't you have a book to write, Dad?"—reminding me regularly that not only do I not know everything, I am still learning every day of my life.

About the Author

Mark C. Bojeun, PhD, has more than twenty years' experience in software development, project and program management, and developing and managing program management offices (PMOs). For the last nine years, Dr. Bojeun has taught program, project, and risk management for Concepts Integration International (www.ci-2.com), a PMI Registered Education Provider (REP). These courses helped hundreds of Project Management Professional (PMP) candidates achieve their certification and have expanded the body of knowledge on project and program management, extending the value of the certification to government as well as to domestic and international corporations.

He also has extensive experience in providing transformational leadership and strategic support to executive management professionals in the development and implementation of organizational vision, mission, and strategic objectives. Dr. Bojeun currently holds Program Management Professional (PgMP) and Project Management Professional (PMP) certifications from the Project Management Institute (PMI), is a Microsoft Certified Solution Developer (MCSD), and holds an MBA from George Mason University and a PhD in organizational leadership.

Currently an adjunct professor at Strayer University and George Mason University, Dr. Bojeun continues to both instruct students in project and program management skills and also build and lead high-performing teams to exceed objectives on programs around the world. He has assisted corporations in creating their own Enterprise Project Management Offices (EPMOs) and has developed processes and best practices for organizations to leverage in consistently achieving project/program success internally and externally.

1

Introduction

I have just completed reading a fantastic novel by John Grisham. As I close the last page of the novel, I am absolutely amazed that I didn't skip to the end as so many say they do with similar novels; yet that would take away the pleasure of letting the thoughts unfold, the drama build, the tension and love for the characters and imagery in your mind develop. Suddenly you are not rooting for the lawyer, innocent of any crime; you are on the side of the lawyer turned criminal who is now showing the U.S. federal government who is the boss.

Would you ever have picked up a book that said, "Read about how the U.S. Federal government was swindled"? No, that would make us terrorists or un-American. Yet, instead, we start at a point in time on a spectrum learning about a single man, forty-five years old, sincere in his belief that he had committed no crime, now starting the sixth year of his ten-year prison sentence—how unfair! How could we let this happen? An innocent man in prison? No, not us.

And with leadership, my mistake was that I wanted to join the noticeable and remarkable authors who have taught us theories such as "great man" and transformational versus transactional, heroic or "level 5 leadership"—all great works pointing toward that pinnacle of greatness that the true chief executive officer of a Fortune 500 company is. Of course we want to study that man/woman and see how he or she achieved such greatness, don't we? In theory yes, in practice NO.

Now give me just one moment. This is a book on how to lead a program team to success. While you might indeed become the next great CEO or world leader, our focus is on leading a program, achieving those objectives, resolving stakeholders needs, realizing benefits, and minimizing risks. We need to lead this team to greatness; there is time to become that heroic CEO in the future, but right now we have real challenges in front of us. Knowing the process, how do we empower, motivate, and drive our program to success?

And that conclusion is the beginning of this book.

While program management is a set of process areas and knowledge areas producing a set of objectives and realizing specified benefits while employing governance over multiple projects and operational activities simultaneously working toward a common goal, it is also the active employment of situational leadership tailored to the program objectives, project manager skill set, operational management knowledge, subject matter expertise, size, duration, and organizational challenges.

We can teach the science and the specific skill sets of program management, but the real challenge is how we drive, motivate, and empower our team toward success. This is not a science, or a one-size-fits-all approach; rather, it is customized, leveraging the knowledge, subject matter expertise, and experience that we have achieved along with all of the leadership approaches we can muster today.

I personally started my drive toward leadership in the early 1990s. During that time, programmers like me looked at managers as obstacles to overcome, not people who were there to assist or even drive our success. In one particular situation, we actually chose the worst of us to become the team manager so that he could do no harm to the program we were writing.

Instead of looking at management as an obstacle, I started my journey to determine how management could be used to help overcome the obstacles and challenges that the organization presented to the delivery of our programs. And along the way, I came to the realization that the challenge was not overcoming the organization so that we could do the work; it was that we were doing work that was not in line with the management objectives or strategies. We were solving individual problems but not working as a cohesive team to support the organization achieve its objectives.

Sure, looking back we were simply dumb; today we can look at the Project Management Institute (PMI®), IT Infrastructure Library (ITIL), and innumerable other standards and models that show how engineers and information technology (IT) support have failed to support organizations in the achievement of strategic objectives regardless of the level of skills employed. Yet even with those tools at our disposal today, we can look at the failures of program managers and program teams to achieve the dreams of the executive leadership. The news is littered with program failures such as:

- The FBI's (2005) Virtual Case File (VCF) Program ($170 million wasted) because of:
  - Repeated management turnover
  - Micromanaging of software developers
  - Poor software-engineering processes
  - Lack of architecture/blueprint
  - Requirement changes
  - Scope creep
- Boeing's (2013) "Dreamliner" being grounded just after the first fleet had been sent into the skies because of electrical fires that were due to:
  - Engineering failures
  - Poor program leadership
  - Known risks

Even the original *Mars Polar Lander* (December 1999) effort—with a $193 million cost of development, $91.7 in launch costs, and $42.8 in mission operations—failed. While the actual reasons for the failure are unknown, the independent review of the program determined that some of the underlying causes of the failure were poor program management and inadequate funding.

Program management itself is not a guarantee of success; it is the leadership and the ability of the team to trust in the wisdom and the team leadership.

This book is not about the leadership style that I leveraged as a CEO or the manner in which I teach courses and seminars on program/project management. Instead the focus of this book is how a leader achieves success with programs. While I do have extensive experience in project management and program management, I have had the opportunity to build a project management office (PMO) or center of excellence for four different companies, all with a somewhat unique set of problems and challenges but all seeking the same set of benefits. Each organization wanted to achieve more consistent success, deliver projects more efficiently, increase overall quality, enforce more effective governance models, and reduce costs, subsequently increasing productivity and profitability.

With such a varied experience set, I have found myself in a number of environments and cultures that required adjusting my personal working styles to the environment, learning new communication patterns, retraining (both employees and executives), building governance models, and reengineering process approaches. Managing programs and projects was simply not enough; leadership and the development of high-performing teams (HPTs) were always the keys to success. Each of these efforts required a transformational leader who demonstrated in-depth working knowledge of the firm, technology, industry best practices, and process. In addition, leaders had to empower staff, drive innovation, resolve conflicts in healthy ways, and demonstrate a level of confidence in themselves, the team, the process, the technology, the vision, and the successful outcome regardless of the risks of the effort. Each situation required a unique set of skills and talents, all leveraging the program management framework and processes, as well as leadership approaches, modified to fit the situation and conformed to work with the unique challenges presented by the team.

SUMMARY OF BOOK

This book is focused on individuals who have come to understand the values of the tools that are provided by PMI but are still looking for the advantages and the success factors necessary to be truly great and effective program and project managers. It is intended for managers and leaders who continue to focus on self-actualization and continuous improvement as a way of learning from historical efforts and driving each new initiative to the highest possible set of standards.

The following pages are intended to cater to managers who believe in the standards of project management and follow the day-to-day mechanisms taught by PMI but who are always striving to achieve a greater value for the team and the organization. They already recognize the gap in leadership tactics and are searching for more answers. Many of the readers will be those who have had extensive experience with executive management, offshoring, geographical diversity, and internal process-based challenges. Most readers will have dealt with geographically diverse teams that can be spread throughout the country or the world and will have endured additional challenges not encountered by collocated team members.

The program manager practitioners, including day-to-day tactical program managers who interact regularly with team members, will be challenged by obstacles such as scope variations, cost issues, technology problems, and outsourcing needs versus the costs of hiring staff (full-time employees). This book will provide the practitioner valuable strategies that are supported by academic research and practical experience.

There are two books required for program management: *A Guide to the Project Management Body of Knowledge* (*PMBOK® Guide*, 3rd ed.) and *The Standard for Program Management*. These two books provide the framework for program management and the formalized steps of program and project management that managers use to achieve success. For formalized process and standards, we look to PMI® as the premier professional organization for project and program management. The primary source of information from PMI® generally stems from the *PMBOK*, but program managers leverage the *PMBOK Guide*, 5th ed., the *Standard for Program Management*, and the Portfolio Standard as guides to processes and tools that can be used for smooth program efforts.

With these guides, a wonderful set of tools/processes for program and project managers is provided to mechanically manage efforts and to achieve program success, but they lack substantive information on the complexity in creating HPTs, individual motivators, and personal agendas brought to every project. To be completely accurate, PMI does reference team development in the human resource knowledge area (much more so in the fifth edition of the book), but still falls short in directly addressing the role that effective leadership and leadership traits play in achieving success.

Quite often I hear that "my project is different" or "I am not doing IT work so these processes do not apply." The *Standard for Program Management*, 3rd edition, and the *PMBOK*, 5th edition, were both authored by volunteers, many of whom are managers from a number of industries including construction, road repair, bridge building, shipbuilding, weapons building, IT software, and networking efforts. Furthermore, PMI's *PMBOK* definition of a project is: "a temporary endeavor undertaken to create a unique product, service or result." A project can create:

- A product that can either be a component of another item, an enhancement of an item, or an end item itself;
- A service or capability to perform a service (e.g., a business function that supports production or distribution);
- An improvement in the existing product or service lines (e.g., A Six Sigma project undertaken to reduce defects); or
- A result such as an outcome or document (e.g., a research project that develops knowledge that can be used to determine if a trend exists or a new process will benefit society). (PMI 2013a, 2)

Projects are:

- Performed by people;
- Constrained by limited resources; and
- Planned, executed and controlled.
- Managed from a time, scope and quality basis
- Projects and operations differ primarily in that operations are ongoing and repetitive, while projects are temporary and unique endeavors. (PMI 2009, 12, PMBOK Guide, 4th ed.)

So if your effort is unique and different from what has been done before, has a limited time frame, and consumes resources, it is by definition a project. The fact that your product, service, or result differs from those done by others who do what you do is exactly why the standards were created. These standards are guidelines to assist in building or creating a unique product, service, or result. If the effort is repeating past efforts, it would fall into an operational approach and management style.

Unfortunately, there is no single formula that can be applied to every project or program. Instead it is the manager's responsibility to tailor his or her management style and the process to the needs of the effort. In my experience, I have not worked on a project that required every process and knowledge area identified in the *PMBOK*. Instead I use it as a toolbox full of tools. I pull out the right tool to solve the problem I am facing and configure the tool to meet the standards of the organization and the success factors of the effort.

This book will not focus on the methodology or process of program management nor on the tool sets that PMI has identified in its program management documentation. I completely agree that these are valuable tools, help in making every program more successful, and are well documented in the PMI materials. Tools such as these are a part of a larger toolbox including technological knowledge, subject matter expertise, and organizational awareness and are meant to be used based on the situation and the environment. The size and duration of the project or program will drive which process tools are most effective and identify those that are unnecessary. Process is a vital part of program management and ensures that critical steps are followed such as risk, scheduling, cost, and change management. Each program/project is a unique endeavor and will have unique needs that will not require the entire PMI tool kit.

While PMI does a great job in identifying the tools and process areas of program management, it lacks sufficient information addressing the overall value of a leader in achieving the objectives of the program. A truly effective program manager is one that recognizes the role that leadership and building HPTs can play in achieving program success, motivating staff, driving innovation, limiting risks, and empowering team members to exceed the objectives of the initiative. The intent of this book is to focus on the value that leadership brings to program management and the impact that it has on the team's success, motivation, morale, and willingness to work cohesively.

I personally find it frustrating that many training facilities deliver program and project management test prep (boot camp) training courses, teaching the test but spending very little time on the role that leadership traits and motivating staff play in the delivery of successful projects. These organizations most often focus on the test, the process areas, and the tool sets but do not dwell on the softer side of management such as driving innovation, increasing communication, positive and healthy conflict resolution, or motivating and empowering team members. My intent is to correct this oversight and to add to the body of knowledge by expanding the thought process to include the value of transformational management and situational leadership as some of the approaches that can assist in creating HPTs in difficult or toxic environments. In addition, this book will include a set of case studies that come from real world examples that will help readers to implement both process and leadership approaches in their program management approaches.

Program success is measured in many ways: cost, quality, scope, benefits, timelines, and, of course, client satisfaction. The challenges encountered with each program undertaken are all different and often unique, whether they are stakeholder politics, technical challenges, cost issues, resourcing, or communications related. Whatever the root cause of an issue is, it falls to the program manager to overcome these obstacles and still provide project managers with a clear road to achieving success. Root causes can be technical challenges, organizational makeups, historical experiences, financial constraints, and stakeholder expectations. Standards and consistent process can be quite helpful, but often program managers will leverage various leadership styles, political approaches, and innovation techniques to overcome challenges and create a safe environment in which team members can share their thoughts and ideas.

In addition to process guides and knowledge areas, a number of factors affect program management success. Among those are situational leadership, innovation, and communication; but most importantly, a vast number of professionals will agree that HPTs are crucial for success and that creating and motivating HPTs is one of the key focal points for program managers. Yet while there are quite a few white papers on leadership and HPTs, there are very few tying program management process together with the tools and traits of how to achieve HPTs with situational leadership. The intent of this book is to lay a general foundation for program managers and provide tools, examples, and strategies to build and lead HPTs, thereby increasing the opportunities for program success. We do know that the way a team performs and the ability of that team to overcome obstacles and achieve objectives is directly related to the leadership strategy employed for the team. Before we drill into HPTs, let's take a moment to more effectively understand and agree on the definition of leadership.

Burns (1978) notes that there are more than 150 definitions of leadership and suggests that "leadership is the reciprocal process of mobilizing, by persons with certain motives and values, various economic, political, and other resources, in a context of competition and comfort, in order to realize goals independently or mutually held by both leaders and followers" (425).

Leaders are generally described as people who are intrinsically motivated, are self-managed, have excellent communication skills, and are visionary, empathetic, and naturally charismatic. A leader is someone others choose to follow and support and someone who can get others to set their personal objectives aside to pursue a new goal contributing to a more common objective (Hogan, Curphy, and Hogan 1994).

Leaders combine individual team members into a more comprehensive team that, when working together, can achieve more than individual efforts would be able to. Leaders don't manage or mandate actions or tasks; instead, they motivate and empower staff to identify and complete the work necessary to achieve the established outcome. Teams led through effective leadership minimize risk, transition conflict from negative and unhealthy to positive and innovative, and as a general rule are capable of exceeding expectations through joint efforts, open communications, and clear lines of responsibility. It is through leadership that a vision can be established to ensure that team members understand the outcome, product, or result that the team is trying to produce. This vision is a clear, concise statement, easily understood, and repeated often so that individuals, stakeholders, and teams can work toward a common objective, avoiding the consequences of ambiguity and confusion.

One of the more common misunderstandings around leadership is that it is not management. Management is instructing personnel, timekeeping, governance, or mandating directives and tasks. *Management is short-term, focusing on the bottom line, and does not work with staff with a focus on motivation or empowerment.* On the other hand, regardless of the leadership that we have in an organization, management is also necessary and will never be replaced. Implementing and directing administrative actions, focusing on the bottom line, and short-term visions are necessary for the achievement of long-term objectives.

Leadership is the motivating and building of teams to achieve an established outcome by creating a positive and healthy environment, using communication channels, implementing conflict resolution strategies, team building, and developing clear roles and responsibilities so that teams more effectively work together. Therefore the result of leadership is the positive achievement of initiatives in the pursuit of goals and objectives that are beneficial to the organization's strategic benefits and have a positive impact on stakeholders. The focus on motivation, comfort, and reciprocal process implies that the leader is working with the team rather than directing them. Whereas management focuses on handling complexity, leadership is centered on change and innovation in the organizational environment. Though leadership can complement management, it will never replace it. The manager has a short-term vision looking at administrative actions, focusing on the bottom line, and doing things right, whereas the leader looks long term, innovates, focuses on the vision and strategic objectives, and does the right thing. The leader directs the activities of a group to achieve a common goal. This common goal must be not only understandable and agreed upon but also communicated to the team in such a way as to ensure that it clearly understands the objective and willingly follows the goal.

Ineffective managers build a hostile and troubled environment where conflict is not resolved and blame is often heaped on individuals. Teams working with ineffective managers often are risk intolerant, are unengaged, and avoid innovation rather than drive toward success. A negative environment such as this is rarely addressed because each program is different and reasons for program failure are commonly spread throughout the team. We often find that team members predict failure early in the process, and while they may share that opinion internally, most don't communicate their concerns with management.

So with a clear definition of leadership, the words and outcomes for leadership are focused on the positive. Leaders are often recognized as having most, if not all, of the following traits:

- Charismatic
- Transformational
- Visionary
- Trustworthy
- Courageous
- Confident
- Motivational
- Innovative
- Effective communications
- Driving the empowerment of staff

Just about everyone I meet has a funny story about a really ineffective manager in his or her past. As a matter of fact, most of us have had the *pleasure* of working with a narcissistic, egotistical, uninformed, or less than competent leader who insisted on wasting valuable project time on meaningless questions and discussions. These ineffective leaders are often ridiculed behind their back and find very little support from the team. Because they are not respected or trusted, these managers often became more of a problem than a solution.

Just recently, I walked into a program based in the software development division of an IT department where the director of product management was fired for repeated failures on projects. Although this leader felt that he knew project management, he understood neither project management nor software development and yet was responsible for managing the entire IT division. Unfortunately, because of his position as an officer of the firm, only a few people complained to the company about his bad management style and his abusive and derogatory comments to employees. I mention this only because this created an overall hostile environment where derogatory comments and personal attacks were common. The organization failed to follow commonsense human resource policies in the workplace and did things like sending out dress code directives for women and not men, setting rules for one group that were not applicable to others in the same position. These kinds of derogatory and discriminatory approaches were inflammatory to the staff and increased the hostility level, further decreasing overall morale and leading to reduced communication and distrust among team members.

My personal favorite example of poor management was one where the manager walked into an office of staff members and announced that he no longer even saw people, just resources to be moved around the board at will. As we stood there shocked, a colleague of mine announced that "the work stops now," and she was serious for at least a week! Yes, sometimes reality is more humorous than fiction.

While oftentimes humorous, the impact that a bad manager can really have on an organization begs a number of questions. Where leaders are driven to achieve results greater than individuals alone can achieve, management is focused on doing what is necessary according to the directives that they receive. Empirically, individuals in those scenarios can evaluate their own actions and the general results of peers, but without evidence it remains opinion. Therefore understanding the actual impact of management is vital to understanding what most workers inherently realize: bad managers kill organizations. Thus evaluating and understanding the actual impact that a bad manager can have on the staff is, and should be, something of great value to all of us.

KEY BENEFITS OF THIS BOOK

1. Although we will be examining leadership, the focal point will be leading the team to achieve high-performing status, succeed in difficult situations, and face greater challenges.
2. The goal of an HPT is to achieve team cohesiveness regardless of geographic location or cultural boundaries and to overcome obstacles in communication and generate innovation and positive conflict resolution.
3. The intent is to address the role that leadership can play in motivating team members to exceed expectations, achieve common goals even when those goals may go against their personal objectives, and produce success regardless of the challenges faced.
4. This book will specifically address the value both of communication from the program management level in both listening and outward messaging and of taking action on realistic improvements to process, product, service, and results, understanding that issues such as

conflict must be resolved immediately or one risks negative conflict, which is destructive to the team.

5. This book will help to extend the somewhat limited information provided by PMI to include the human factors, HPT development, and motivation beyond the physical tools that PMI focuses on for project delivery.
6. This book will identify and explain a number of situational leadership styles, the value of each one depending on the organizational culture and environment, and the impact that styles can have on the team members and goals of the program.
7. This book will offer case studies, quizzes, and other material that make this an excellent book for a course on program management or complexity.

Section I

Leadership and Program Management

The task of the leader is to get his people from where they are to where they have not been.

—**Henry Kissinger**

2

Leadership Study

> Leadership is the art of getting someone else to do something you want done because he wants to do it.
>
> **—Dwight D. Eisenhower**

Good leadership drives motivation, innovation, creativity, conflict resolution, and team development while increasing risk tolerance and expanding communication channels. Therefore we can also assume that bad leadership will imply negative consequences for the team, program/project, and strategic objectives for the firm.

To understand the value that leadership can play with successful programs, the best place to start is to understand the consequences of bad management. When bad leadership or management is employed, there can be some serious consequences. Recently, Ericson, Shaw, and Agabe published an article in the *Journal of Leadership Studies* entitled "An Empirical Investigation of the Antecedents, Behaviors, and Outcomes of Bad Leadership" (2007). Although the article focuses on bad leadership approaches and styles, it fits into the general premise of the manager-versus-leader discussion in that it focuses on the outcome of a bad leader regardless of the individual's title. The entire article is worth exploring, but here I address just some key points.

In this study, 335 participants responded in full or in part to the twenty-one-question survey, and although the results are not surprising, they are evidence for empirical impressions established by most professionals. Respondents were asked to focus on a personal experience that they had with what they deemed to be a bad leader and to answer the questions accordingly. Based on this scenario, the study may be slightly skewed but the results are eye-opening.

In the answers provided to this survey, bad leaders were identified as those who had difficulty dealing with subordinates (17.6 percent), poor ethics/integrity (13.3 percent), poor interpersonal skills (11.5 percent), and poor personal skills (14.1 percent). These issues resulted in employees feeling frustrated (11.6 percent), feeling angry (15 percent), and having lowered self-esteem (13.9 percent). In addition, the bad leadership was directly attributable to the development of a bad organizational culture (17.3 percent), overall performance loss (16.0 percent), attrition of employees (21.3 percent), and motivation loss (12.8 percent).

What was surprising and somewhat disheartening was that when asked what happened to the bad leader, 44.8 percent of the participants stated that he or she was either promoted or rewarded and 13.4 percent stated that nothing happened to that individual. Therefore 58.2 percent of the respondents reported a situation where the manager would most likely continue to have a negative impact on the organization and its staff.

With the motivation, performance, human resource, and cultural losses that a bad leader is able to cause in an organization, it is surprising that so many businesses keep or promote bad managers. Does the organization really understand the cost of this leadership and the impact that it has on its organization, or is the focus on the outcomes of the program regardless of the cost, and is the loss of attrition and motivation an acceptable trade-off? It seems not, for even the most ruthless of companies is focused on profits, market share, and growth and recognizes that the loss of performance, motivation, and human resources is an obstacle that must be overcome. In today's increasingly challenging marketplace, if a company actually understood what bad leadership was and the overall impact that it can have on their organization, logic dictates that it would take action and would actively work to eliminate bad management from its corridors. Bad management presents a hefty cost to business in morale, attrition, and productivity loss.

I recently had the opportunity to watch the absolutely worst manager I had ever come across somehow able to hang on to his job for over two and a half years before being demoted. This individual felt that discussion was a waste of time and that everyone should know what he meant regardless of what he said or how he said it. In one case, he told one of his direct reports to go ahead and take a vacation because he needed to get used to being out of the office. The employee took this comment personally and immediately started looking for a new job. Single-handedly, the manager undermined morale, decreased innovation, and increased risks because no one in the organization was comfortable discussing concerns with him. And yet he held on to his job as an officer of the firm, blaming every manager who worked for him when failures occurred.

Therefore the general assumption is that this study is one of the first to start to honestly identify what the cost of bad leadership/management is to an organization. As additional studies are performed and provide more insight into bad leadership traits and associated costs, hopefully this point will be driven home and provide the motivation necessary for upper management to start to remove bad managers and develop honest and valid leadership programs complete with metrics that outline the overall effectiveness of the leader.

Some of the key questions and responses from Ericson, Shaw, and Agabe's (2007) study are highlighted in the following discussion. See reference Table 2.1 for a complete list.

Amazingly, this study found that in 58.2 percent of the cases cited, bad leaders either were not disciplined or were promoted within the organization. Recognizing the impact these leaders had on subordinates' motivation, morale, and stress levels (including impact outside the organization), and the fact that when a leader is perceived to be a "bad leader" by one it is often accepted by many, the long-term effects of bad leadership can have a tremendous impact on organizational culture and environment as well as on team morale and job satisfaction (Ericson, Shaw, and Agabe 2007).

Can you image that over 58.2 percent of the managers were never held responsible for their behavior or inability to achieve? That is what we must overcome. As leaders we must learn that motivation, morale, innovation, positive conflict resolution, and open communication are critical to building HPTs and successful program efforts. These examples demonstrate what we must set as a standard as unacceptable to program managers. And that is the one of the primary purposes of this book, to overcome bad leadership and employ positive collaboration and HPTs.

In program management, leaders are critical to the success process; they coordinate work efforts, identify issues, ensure that consistent management information is communicated clearly and concisely, teach positive conflict resolution, define roles and responsibilities, ensure that everyone understands the goals/benefits/risks of the program effort, and manage the stakeholder expectations, celebrating each benefit as it is achieved.

As you read this book, attempt to identify some of your personal key leadership traits that contribute to being a successful program manager.

TABLE 2.1

Leadership Study

What actions caused you to classify the person as a bad leader?	
Unable to deal with subordinates	17.6%
Poor ethics/integrity	13.3%
Poor interpersonal behavior	11.5%
Poor personal behavior	14.1%
How did the bad leader make you feel?	
Angry	15.0%
Frustrated	11.6%
Lowered self-view/self-esteem	13.9%
What effect did the bad leader have on your work performance?	
Motivation loss	33.6%
What effect did the bad leader have on you personally?	
Negative effect on nonwork life	15.0%
Negative effect	15.9%
Increased my stress	29.3%
What effect did the bad leader have on the organization?	
Created a bad organizational culture	17.3%
Human resource loss	21.3%
Motivation loss	12.8%
Performance loss	16.0%
What happened to the bad leader?	
Bad leader was promoted/rewarded	44.8%
Nothing	13.4%

These are a great place to start as we look at the process and standards of the PMI.

3

Developing and Achieving a Common Vision

The leader is one who mobilizes others toward a goal shared by leaders and followers. . . . Leaders, followers and goals make up the three equally necessary supports for leadership.

—**Gary Wills**
Certain Trumpets: The Call of Leaders

How often have you heard the phrase "I am like a mushroom, kept in the dark and fed BS"? This phrase is one of frustration from team members identifying the concern that they do not know what they are working on, why it is important, or how it will contribute to success. Often they are not even aware of the components that must be integrated to achieve the program objectives. It has been proven time and again through numerous examples of failed projects that compartmentalizing people and providing limited information leaves them unsure of their contribution and unable to take advantage of opportunities available to them by understanding the overall vision of the effort. Keeping team members in the dark increases risk for both threats and opportunities and creates a scenario of distrust between team members.

Informed team members understand the end game and become aware of efforts outside of their area of responsibility. Through this information gathering, teams can gain insight into the overall program and begin to identify areas of redundancies, reuse, and risks. In a positive environment, team members are open about the risks they identify in the project and can establish mitigation strategies for it and potentially include it in the expected monetary risk assessment. In addition, it is crucial to understand that a negative risk in one project could be a positive risk in another project or in the program as a whole. When team members communicate these risks, it enables the program manager and their project managers to be aware of the risks and to possibly leverage them into positives for the overall effort. With awareness and knowledge, teams can work together to ensure a clear and concise view of the product, service, or result, enabling individuals to understand how their part of the program will contribute to the overall success of the effort. We as program managers need to include the team in the development of the vision, risks, and benefit realization plans. Team members responsible for QA, development, engineering, marketing, customer service, and requirements all can offer helpful suggestions about the challenges, opportunities, and land mines we will encounter as we take on the effort. I believe in involving the team as early as possible to ensure that we don't waste time chasing down issues that are unnecessary to the program or project rather than identifying a viable solution to the problem.

On multiple programs, I have encountered situations where the current state of the program was anywhere from the initiating stage to mid-program, only to be surprised that many of the team members either had no idea of what the program was supposed to produce, had conflicting opinions of the program intention, or did not know the scope that was intended. In some cases, program team members were uninformed as to what their contribution to the overall effort was and why their contribution was critical to the success of the effort. Based on this lack of information, team members would struggle to identify necessary requirements or understand how their work affected the critical path. Obviously, there was no way to know if the effort they were working on had additional opportunities to contribute to program objectives.

You would be amazed at how often program managers start an effort without a clear understanding of the program, complexity, risks, assumptions, and constraints. When assigned a program, the very first question to be asked should be, "What does success look like, and how will we know when we have achieved it?" Unfortunately, many stakeholders do not know. This is a process that a program manager needs to go through to identify what the goals are—not just the high level, but the detailed objectives and how they will affect the stakeholder community and benefit the organization at large.

Understanding what success looks like is the first step in developing a vision for the program. We use the term "vision" as a clear and concise statement to describe the characteristics and features of what the final product or service will be. It is a complex set of objectives, goals, and benefits that the program will deliver and is aligned with the strategic objectives of the organization and reflective of the strategic direction established by the executive leadership team. The vision will define what will be produced, how it will operate, and what process it will execute, as well as what is not included in the program (out of scope).

Because programs are the creation of a new result, product, or service through the culmination of multiple projects, operational efforts, and process change, program vision is crucial to the success of a program, but it is also the most common reason for program failure. If the vision of the product, service, or result is not clear, time and energy can be wasted on work that is unnecessary to the finished effort. It is impossible to build something we can't envision, and managing the expectations of the stakeholders and sponsors is not something that can be accomplished if we don't all agree on what the product, service, or result characteristics should entail. When success cannot be defined, there is no way to establish and realize benefits from the effort. Furthermore, without a clear vision programs can deviate from the organizational strategy. Developing a clear and concise vision that is understood and shared by all team members is the direct responsibility of the program manager and must be approved and supported by the program sponsor.

A program manager will use the program vision to ensure that everyone understands what the effort is intended to produce. Program managers must be able to be both a sender and a listener in communicating the vision. In other words, they cannot simply state the vision and expect understanding; they need to elicit questions, concerns, and assumptions, and to gather recommendations. This conversation is one that must take place throughout the program life cycle, and it is the program manager's duty to be willing to adjust as new information is identified. We will discuss conversational approaches later in the book, but realize that the sender-listener relationship is one way that program managers build confidence and trust in their team members.

There are quite a few times that I have found the need to adjust program scope or the benefits realization plan because when we hit technical or usability issues, there were better solutions available to us, or ideas evolved that offered additional opportunities to the program. So the conversation around vision is an ongoing one. And it is through the vision that we can better understand the benefits that a program will achieve and the timing of those benefits, and therefore realize the benefits for the organizational stakeholders.

On the other side of the coin, program managers can also leverage the program vision to decrease or bind the outcome (scope) of a program and to ensure that benefits not included in the program are communicated and understood by team members and stakeholders to more effectively manage expectations around program outcomes. When a program vision is not clearly communicated, stakeholders may have expectations that are not communicated and will be disappointed when the final product, service, or result does not meet their desire or needs. A successful program manager will ensure that the benefits of the program are clearly communicated and that the factors outside of scope are clearly understood by all.

A program vision will be a technical or strategic road map that is derived from the benefits realization plan that will be achieved through the life cycle of the program. Individual projects will be started and completed during the life cycle of the program, and while each will achieve benefits, it is the culmination of all of the projects that achieve the complete set of benefits that the program is intended to deliver. As such, when a project is completed and reaches closure, its operational control may stay under the program until other projects are completed and operational control can be effectively transitioned to the end user, client, or functional operational departments.

A clear, concise, and nonambiguous understanding of what the effort is intended to create and why that is beneficial to the firm or customer is required to ensure that both program stakeholders and team members know what is being built and how the program will be beneficial to them. Each team member should completely understand what is required, who is doing what portion of the program, and not only what he or she is working on individually but also on why it is important to the program and the effort as a whole.

In addition, if we don't know what we are building, how will we ever know when we are done or what success looks like? Understanding the vision provides us with decision-making boundaries that are used in everyday actions and enables the team members to ensure that a common goal is followed for the program at hand. However, if the vision is even slightly off, project team members can end up with gold-plated features, adding additional functionality that is unnecessary, or miss crucial aspects of the feature sets. As previously mentioned, a vision should be clear, concise, and leave no room for ambiguity. Something like "software to enhance the user experience" or "create a service that optimizes the customer satisfaction levels" can be left open to interpretation and lead to decisions that are well outside the scope of the program. Instead a vision should ensure that anyone who reads it or hears it will have the same understanding. A better program vision would be: "Eliminate current road congestion problems by turning all stoplights on Highway 28 into overpasses from Highway 66 to Highway 267 decreasing road congestion by 60%."

That is not to say that any one vision or even the program manager's vision is the correct one. Quite often the final vision for a product will differ dramatically from what the initial vision looked like and will include contributions from all team members and stakeholders. Enhancements requested, opportunities found, and learning will all contribute to the vision as it evolves through the life of the program. Because of the evolution of a program vision, it is critical that not only does everyone understand what the vision is; they also must be informed as the vision evolves and team buy-in is achieved. As has been mentioned, if the team does not understand the evolution of the vision or disagrees with the choices, the delivery of result, product, or service will suffer. Changes to what the product or service will look like or do will require team members to make adjustments to their efforts, ensuring that they are staying in line with the end goal.

I worked on one program that was specifically designed for an expert set of researchers to be able to better do their searching of intellectual property. However, as the program advanced we found that by eliminating the search key commands and instead using common language for searching, engineers could add to their knowledge and work by searching themselves. This dramatically reduced the number of failed requests for patents, because the engineer could see that someone had already filed a similar or same idea and could then develop nuances or changes that made his or her approach unique in the field. The overall program included software development, marketing, data conversions, data centers located worldwide, and sales efforts.

Because the software development aspect was only a piece of the overall program and had a unique product that it would produce with limited time and resources, it was defined as a project within the program and required a clearly defined vision. A project vision is created from the overall program vision to define the unique needs of the project and how it will continue to the program's benefit. In many cases, minor differences of opinions on the project's vision can create chaos for the overall effort and result in costly rework. In other situations, the project can be canceled as it no longer meets the needs of the program. For example, a project that was estimated to require more than a year to build with the costs consuming too much of the program budget could be outside the scope of the program itself, and the project would be canceled. The program would have to reevaluate the needs and define a new project that would achieve the benefits desired without the overruns of cost and schedule. If this were a software development effort, the decision could result in a make-versus-buy decision, and the costs of the purchasing of commercial off-the-shelf could be less than the costs of building a new software application.

The program vision sets the stage, and the project vision aligns with the program so that the defined benefits related to the project can be achieved. Every project team member must have the same understanding of what the program and project will result in, who the stakeholders are, and how the project will result in achieving the defined benefits for the program. While the program may have additional benefits that other projects and operational efforts will achieve, the project contributes to these benefits, and understanding the contribution is necessary for the project team to realize the benefits. If this is not the case, minor decisions made by project participants can result in costly rework and scope creep. Had the information been readily shared, opportunities for reuse and eliminating redundancies could be leveraged.

On one program that I was asked to take over, the effort to create an online software application had been going on for six months, and the client was so unhappy with the progress that they canceled the program. At that time, the team consisted of five contracting companies, located around the globe, and almost 100 total team members. I was asked to take over and restart the program. The very first thing I did was to travel to each of the organizations involved in the program and ask them to tell me what the program was supposed to look like in its end state.

Each of the teams, and some individuals within teams, presented differing and sometimes conflicting views on the end state. This was not a question of the client and contractor having different views based on their perspectives but, instead, contractors and development companies conflicting externally and internally on the vital aspect of what the program was intended to be. I was left with the distinct view that if we cannot envision what we are building, how can we possibly build it?

When I returned from my site visits, with a very confused view, I sat down with the client to understand its desire for the effort and, amazingly enough, I heard yet another vision for the program. Obviously, the policies, procedures, documentation, and all the formal meetings were meaningless if the teams did not agree on and understand what they were building.

The failure of the program was not the methodology or the conformity to a set of processes; instead it was the leadership of the program and the inability of the program manager to establish and communicate a common and accepted vision motivating disparate parties to all contribute toward a common goal. Once we overcame this hurdle, the process and development of the product ran incredibly smoothly.

The exercise was painful, and I admit that the vision had to be adjusted for some very adamant individuals who did not want to change their view; but eventually everyone was on the same page as to what the program benefits would look like and the product it would offer. At the end of the day, the vision for the program was something everyone felt ownership in because they contributed to it and were vested in the success of implementing an enterprise-wide vision. Each team member could point to a specific feature or function that he or she contributed to the vision and as such took ownership of the program personally, making it successful. Once we had the vision established, the project was able to get back on track with every member working cohesively with other team members, contributing to a master schedule and starting to form into an HPT.

Although this team was a conglomeration of over 5 outside different consulting firms and one client, a Microsoft executive who joined a team lunch commented that he could not differentiate who worked for which company as the entire team intermixed with open conversation on the challenges and obstacles they faced. This HPT was observed offering suggestions for other team members to problems that have not been resolved.

It is important to recognize that the PMI framework for program and project management was implemented, and the teams generally operated with a formalized development approach; but without leadership and a common vision, success was simply not possible. Each product elaboration had to overcome the visionary issues and address concerns that should have been identified and handled at the program onset. Many hurdles were ahead, but with the clearly defined vision and established roles and responsibilities documented, the teams were able to dispense with meaningless infighting and start focusing on developing a solution that not only was on time and on budget but also exceeded client expectations.

Of course, the vision had to be repeated clearly and concisely often throughout the program life cycle. Without this, team members could get off track, identifying potential new ideas and gold plating the program. The goal of the program manager was to stick with the vision, communicating clearly and often so that any confusion as to what was being built and why it would be beneficial to the organization could be eliminated.

The achievement of the program's defined benefits allows the organization to reap the reform they anticipated from their very large investment in time and money.

4

The History of Project and Program Management

PROJECT MANAGEMENT HISTORY

The very best place to start in understanding program and project management is to have a comprehensive knowledge of PMI-based definitions for project, programs, and the management process used to successfully achieve objectives. Project management is the foundation of program management, and program managers will generally be experts in the field of project management to ensure that the project managers are following a consistent and repeatable process.

The PMI states: "A project is a temporary endeavor undertaken to create a unique product, service or result" (PMI 2009, 4). It is:

- Performed by people;
- Constrained by limited resources; and
- Planned, executed and controlled.
- Managed from a time, scope and quality basis.
- Projects and operations differ primarily in that operations are ongoing and repetitive, while projects are temporary and unique endeavors.

While PMI does a great job describing a project, many in the industry would describe project management as "the art of creating the illusion that any outcome is the result of a series of predetermined, deliberate acts when, in fact, it was dumb luck" (Anonymous). Project management is best defined as "the application of knowledge, skills, tools, techniques to project activities to meet project requirements." (PMI 2009, 6).

Contrary to popular belief, project management is not related to "dumb luck" nor is it an IT or software development concept. Project management, as it is reflected today, really began in the 1950s when businesses started to focus on the successful delivery of products through better organization spanning multiple functional areas. These process efforts were mainly focused on better communications and integration of processes across functional borders.

However, project management can be traced further back, for example, to the original building of the transcontinental railroad in the early 1870s. The scope of organizing such a complex effort across the country with limited communication and thousands of workers needed to be optimized to facilitate cost-effective measures and time factors. In addition, the manufacturing and assembly of large quantities of raw materials required a tremendous amount of logistics planning to ensure that workers had the materials they needed available to them as they built the railroad lines. If too many materials were shipped to a location, the cost of transporting materials increased as the materials had to be hauled from location to location. If there were not enough materials, the workers would be delayed and would have to wait, doing nothing and being paid. The balancing act between logistics and management was one of the biggest challenges that managers had faced and required unique management approaches.

As the railroads were being built, Frederick Taylor (1856–1915), also known as "the father of scientific management," began his study of work processes by leveraging scientific reasoning to work efforts, showing that labor can be analyzed and improved by decomposing the process to its elementary parts. Taylor looked at productivity studies in places such as steel mills. Through the use of scientific reasoning, he optimized the approaches, applying his tactics to tasks such as shoveling sand and lifting and moving parts to determine more efficient and effective processes. Instead of just increasing the number of hours worked by the mill workers, Taylor determined ways to optimize efficiency and increase productivity by eliminating overtime and additional staffing needs through the optimization of work processes.

During the same time frame, Henry Gantt (1861–1919) studied the order of work and the processes that manufacturing facilities followed to increase efficiency in working processes. Gantt focused on the construction of naval vessels during World War I. By approaching the work process by leveraging charts containing task bars and milestone markers, he was able to provide a better perspective on the sequence, duration, and allocation of resources. These charts were so useful to management that they were used in the original format until the 1990s when links were added to show critical path methods (CPMs) and the assigned resources, order of events, and

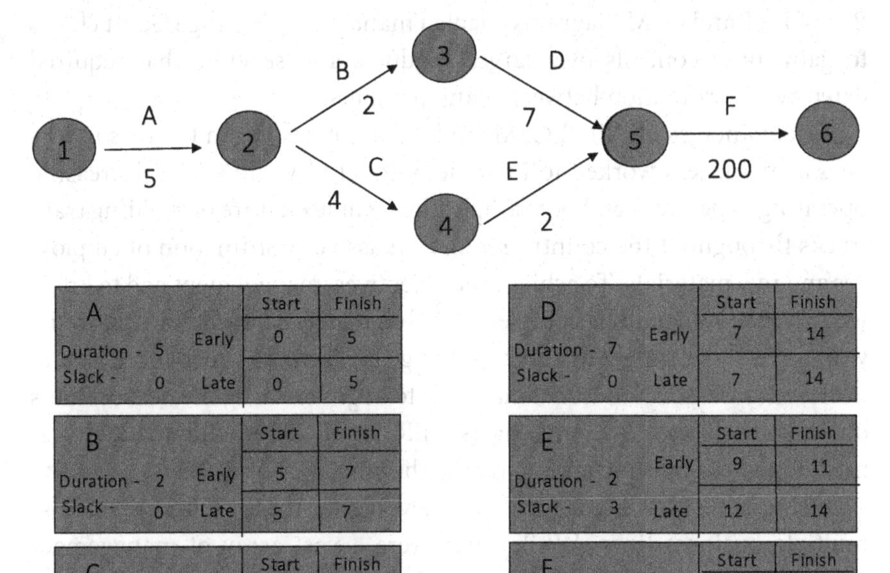

sequencing that were necessary for project completion. These links showed the precedence and relationships between tasks, facilitating a better understanding of resource allocation and enabling managers to visually represent the tasks, milestones, and deliverables on a calendar basis.

PERT (Program Evaluation Review Technique) charts and the CPM were introduced after World War II when the complexities of processes and competition increased but the demand from wartime decreased. Managers needed to optimize and increase efficiency, delivering on time

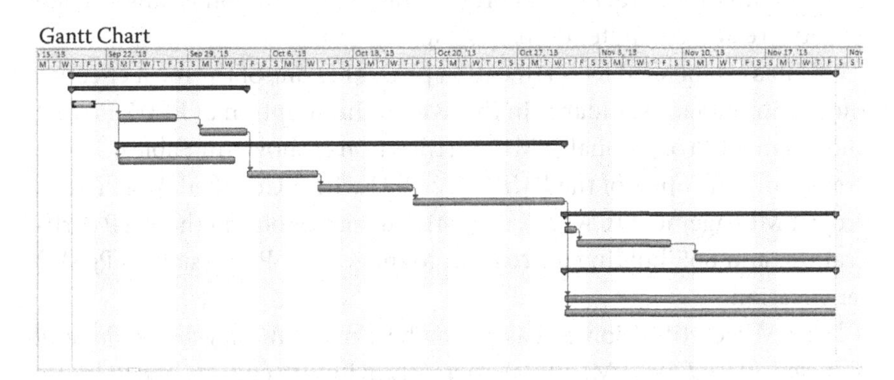

and on budget on a greater scale to meet demands and remain profitable. Both PERT and CPM diagrams enabled managers to leverage Gantt charts to gain more controls over larger products and services that required extensive coordination between team members.

The advantages of PERT, CPM, and Gantt charts began to cross industries as managers worked to increase productivity while also decreasing operating expenditures. For example, the complex nature of building train tracks throughout the country required massive coordination of employees and raw materials. To achieve the objectives, management had to know when a section would be completed so that materials would be sent to the work crews in a timely manner enabling the crews to move forward and not have to wait on additional materials to begin creating the next sections of track. Tools such as Gantt charts, PERT, and CPM facilitated management logistics planning and increased the efficiency of workers.

During the 1960s managers began leveraging these techniques across multiple business lines. While there were a vast array of management approaches that were used, almost all shared the commonality of a leader (project manager) and a team with the focus on communication and integration of team members to optimize work flow across departmental lines and work toward a common vision.

Although there were a tremendous number of organizations that attempted to formalize the process of project management, the PMI, established in 1969, is the industry standard for standards and certification for project managers. With over 500,000 members in more than 171 countries, the PMI identifies itself as "the leading membership association for the Project Management profession" (www.pmi.org). PMI focuses on the development and adoption of professional standards, promotes the development of uniform process, and serves as a forum for project managers to identify best practices on projects ranging from home construction to software and computer development.

PMI has established its certification process as one of the most effective and commonplace standards in the world. The adoption of PMI's Project Management Professional (PMP) certification is shown in Table 4.1 with over 4 million copies of the *PMBOK*, 20,993 with a Certified Associate in Project Management (CAPM), 525,341 currently holding the PMP certification, and 865 holding a Program Management Professional (PgMP) certification.

The PMP certification is based on the *Project Management Body of Knowledge* (*PMBOK*, PMI 2013a), currently in its fifth edition. The *PMBOK*,

TABLE 4.1

Project Management Professionals

Credentials/Certifications as of May 2013	
Certified Associate in Project Management (CAPM)	20,993
Project Management Professional (PMP)	525,341
Program Management Professional (PgMP)	865
PMI Risk Management Professional (PMI-RMP)	1,969
PMI Schedule Management Professional (PMI-SP)	871
PMI Agile Certified Professional (PMI-ACP)	2,635

fifth edition, provides a set of five process areas and ten knowledge areas that demonstrate forty-seven processes to be used by project managers in the delivery of results, products, or services. The *PMBOK Guide* is organized by both process and knowledge areas. The processes interact and overlap within a project's various phases. For any process, three parts are necessary—inputs, tools and techniques, and outputs. Specifically, inputs refer to documents, plans, and designs; tools and techniques are those mechanisms that are applied to the inputs; outputs may be documents or products as well as other types of project results.

The five processes are (see Figure 4.1):

1. Initiating
2. Planning
3. Executing
4. Monitoring and Controlling
5. Closing

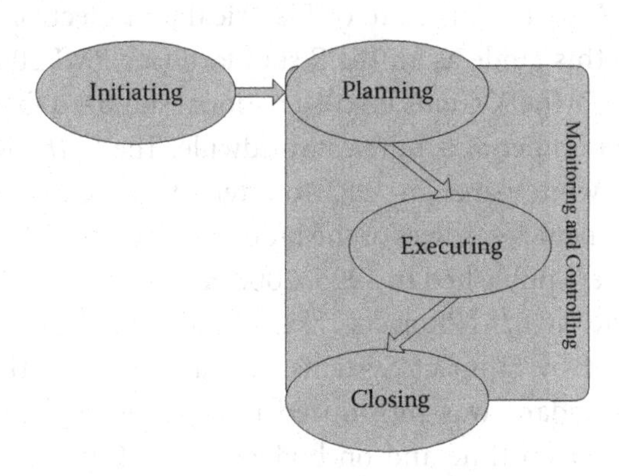

FIGURE 4.1
Project life cycle.

And the ten knowledge areas are:

1. Project Integration Management
2. Project Scope Management
3. Project Time Management
4. Project Cost Management
5. Project Quality Management
6. Project Human Resource Management
7. Project Communications Management
8. Project Risk Management
9. Project Procurement Management
10. Stakeholder Management

As stated previously, these five process and ten knowledge areas combine to provide all forty-seven processes that are part of the project management practice. When combined they provide a framework for project managers to develop strategies for project success. A table on page 61 of the *PMBOK*, fifth edition (PMI 2013a), shows how these process and knowledge areas combine into a toolbox that project managers can use to customize their process for the needs of the specific project at hand. Very rarely are all of the forty-seven processes put in place. Instead, the processes are leveraged where appropriate, and to keep the project manageable, only those that are useful are implemented—the process and knowledge areas as well as each step in the knowledge area.

The *PMBOK Guide* (PMI 2004) was approved by the American National Standard Institute as the national project management standard for the United States. Also, the Institute of Electrical and Electronics Engineers (IEEE) accepts this guide as an IEEE standard (see PMI 2005). It is noteworthy that the *PMBOK Guide* is still a national standard that is advocated and in use in a number of countries worldwide. The *PMBOK* was initially developed as a white paper in 1983 to attempt to identify standards for project management regardless of the industry. The current five editions of this standard were published in 1996, 2000, and 2004, respectively.

In Europe, PRINCE2® (PRojects IN Controlled Environments) was initially introduced by Simpact Systems Ltd. as the PROMPTII standard in 1975. The standard was introduced because of increased failures in projects to deliver on time and on budget. In 1989 PRINCE was implemented through the English government agency Central Computer and Telecommunications Agency (CCTA) as a standard for IT-related efforts.

PRINCE became the de facto standard for all government-related projects. In 1996 PRINCE was upgraded to PRINCE2 and was contributed to by over 150 organizations to achieve a set of standards that met more of the needs of all projects from IT related to construction and electrical engineering.

Finally, in 2009 PRINCE2 was upgraded to increase its efficiency by leveraging seven basic principles of management:

1. Business case
2. Organization
3. Plans
4. Risk
5. Progress
6. Quality
7. Issues and change

While PRINCE2 has a large following in Europe and the UK, PMI standards are also being accepted and are often considered interchangeable with PMI's *PMBOK*. While very similar in nature, PRINCE2 offers a methodology for managing projects whereas PMI provides guidelines.

PROGRAM MANAGEMENT HISTORY

Program management finds its very same roots in the work that Gantt and Taylor did in relation to the scientific study of work processes. Although program management has evolved more recently into a formalized process, it contains multiple projects and can cross into operational management where necessary. Programs are also generally much longer in duration and consist of larger systems, while some are smaller programs benefiting from centralized management of multiple projects. As of last count, there were only 1,000 PgMPs. This may be due to the higher cost of the PgMP certificate ($1,500), an in-depth MRA assessment, or because the test is based on understanding the *PMBOK* as well as Program Management standards.

PMI Definition of a Program

Program management is the process of managing several related projects. "A program is a group of related projects managed in a coordinated way to obtain benefits and control not available from managing them individually" (PMI 2013b, 4).

Programs:

- May include related work outside of the scope of the discrete projects in the programs
- Program components can include the projects as well as:
    - Infrastructure
    - Management effort
    - Operational tasks

Program management is the centralized coordinated management of a program to achieve the program's strategic objectives and benefits that cannot be achieved through a single project. It involves aligning multiple projects to achieve the program goals and allows for optimized or integrated cost, schedule, and effort. Projects in a program have a common or complementary deliverable or capability (PMI 2013b, 8).

Program management is responsible for overseeing all projects in a program and leverages the oversight to support project-level activities to ensure the overall program goals and objectives. The program manager is responsible for trade-offs and clearing project roadblocks, occasionally choosing program priorities to make adjustments to resource allocations between projects. Program managers are responsible for each project's deliverables and have a higher-level view of the strategic goals and benefits the program needs to produce to remain in alignment with the organization's strategic objectives. As such, a program manager can see across multiple projects to be able to better determine strategies, and resource allocations from a personnel and budgetary viewpoint.

A program manager oversees the cross-project dependencies and program-level risks and opportunities, and ensures that governance takes place with each project following a defined process that provides financial, governance, metrics, dashboard, stakeholder communication, and schedule information in a timely manner. Through this, the program manager provides insight for upper management and ensures that the program benefits are achieved in an effective, timely and efficient manner. The program manager has a much higher view of the program and can avoid the silos that project managers often find themselves in with regard to project resources and priorities. Whereas a project manager is focused on delivering project-specific goals on a somewhat myopic level, the program manager must meet the expectations and benefits for the program as a whole. The stakeholder community for a project is those people who are affected by the project outcomes, but a program manager is responsible for the communications with all stakeholders across all projects.

While many organizations consider program management to be the managing of a large project with subprojects being managed by individual project managers, PMI's definition of a program is a clear and concise management approach that provides a structure directly aligned with organizational objectives and strategies. Program management follows five main domains as an approach: (1) strategy alignment, (2) program benefits management, (3) program stakeholder engagement, (4) program governance, and (5) program life cycle management. These five domains permeate everything a program manager is responsible for, from ensuring that expectations are clearly communicated to delivering benefits on time and in a cost-effective manner.

To be truly effective as a program or project manager, you need to be able to build and manage HPTs that produce results efficiently and effectively. As a general rule, the leadership provided by the program manager will determine the environment for projects and can directly affect how effectively the teams operate, communicate, respond to crises, make decisions, and drive innovation. Teams will be collocated or geographically diverse, and the leadership style, approach, and investment will differ depending on the environment, team history, and experience levels. Teams that have already established communications and have worked successfully together in the past may be more empowered and capable of operating autonomously, while teams who have not worked together and operate in a geographically diverse environment will require more hands-on management, team building, communication, and strategies designed to improve team performance.

OVERVIEW OF PROGRAM MANAGEMENT

As stated previously, five main domains are defined for program management:

1. Strategy alignment
2. Program benefits management

3. Program stakeholder engagement
4. Program governance
5. Program life cycle management

These five domains drive the decision making and process for program management to ensure that program managers focus on the high-level deliverables and the leveraging of resources across projects, operations, and the program's high-level benefits.

Strategy Alignment

Strategy alignment (see Figure 4.2) is the process through which the program is initially and continuously aligned with the defined strategic objectives of the organization. Programs contribute to the achievement of organizational objectives through the delivery of benefits. As the benefit delivery is established, the program alignment should have been performed to ensure that the program is contributing to the strategic objectives and that the contribution is tangible and measurable. A program aligned with the strategic initiatives of the organization will contribute to the achievement of the organizational goals and will help the organization to move in the direction deemed most effective by the executive management team.

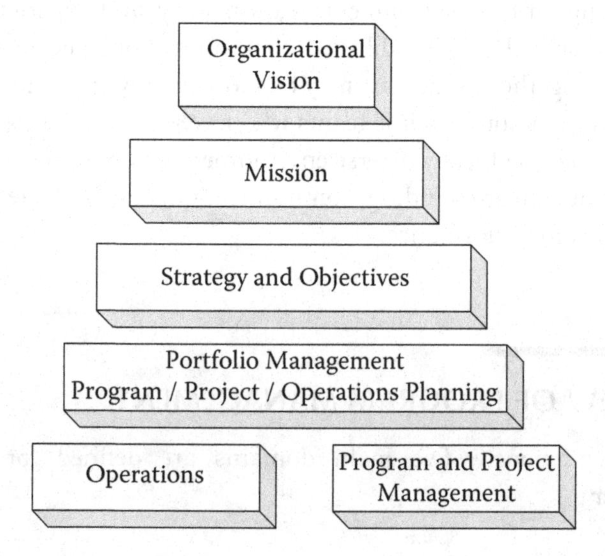

FIGURE 4.2
Strategic alignment.

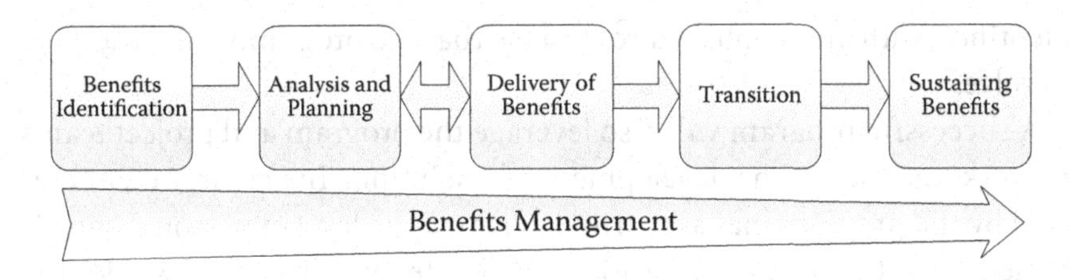

FIGURE 4.3
Benefits management.

Program Benefit Management

Benefit management, illustrated in Figure 4.3, is the identification, realization, and communication of tangible benefits that the program will deliver or has delivered to the stakeholder community. The benefits of a program are most likely to be lost or forgotten with time and the general impact of delivery if they are not regularly communicated and validated. It is the sustaining of long-term benefits and the realization of these benefits that the program manager is directly responsible for. The program manager is ultimately responsible for ensuring that the benefits are useful, aligned with the organizational objectives, communicated throughout the organization, and when realized, validated to ensure that the benefits met the-needs as intended.

Therefore the program manager must be capable of envisioning the future state, providing a gap analysis between the current and future, and ensuring that stakeholders and program team members all recognize and see the value of the future state. A program that is undertaken with no tangible benefits has little support from the organization and will often be challenged on the basis of cost, organizational impact, and resource consumption. A program with a large number of benefits will receive a vast amount of support and resources to ensure that it is able to achieve established goals. At the same time, the greater the benefits the more pressure to deliver in a timely manner and to require communication of what has been realized and what benefits are still pending.

For any and all efforts, the program manager must be able to step up as a leader to the community and to be able to align pain points with benefits so that stakeholders can recognize that the future state will be a better one and are prepared to deal with any temporary pain points while waiting for the new set of benefits that will be an outcome of the effort. The pain points that the stakeholder community experiences must be real and identified with an established root cause that the program will undertake to solve.

A successful program will also leverage the program and project teams to work together in the development of a solution. The environment created by the program manager will determine the level of personal contribution, investment, and innovation that the team delivers as a whole. The program manager must ensure that every team member clearly understands the benefits of the program and the future state, and is on the same page working toward the same vision. Without this, teams will head in disparate and sometimes opposite directions, burning time, resources, and dollars without tangible benefits. Thus, as the responsible party for benefit management, not only must the program manager define the environment, but he or she also must be someone who can envision the future and must be an exceptional communicator to ensure that everyone involved with the program, whether a contributor or user, is following the same vision and overall expectations.

Stakeholders need to have clarity in what will be forthcoming, when things will occur, and when they will receive information. Without this, the program soon becomes a "black box," which is often resented by those waiting for benefits. Regardless of the technical or managerial knowledge of the stakeholder community, there is an expectation that they will be capable of observing some level of progress. Often this progress can be demonstrated through tangible achievements, but there are times, especially early on in a program, where the progress is more intangible. The program manager must be able to effectively communicate progress and schedule throughout each stage of the effort. Stakeholders need to know not only what the status is but when they will hear additional status updates and the form that communication will take. In the PMI standard (PMI 2013b), this is accomplished through the communication plan; but it also is anticipated that the program manager will be able to respond to questions, deliver status on a moment's notice, and recognize when the communication schedule needs to be adjusted to better meet the needs of the community.

Program Stakeholder Engagement

Stakeholders are a crucial element to the success of a program and are not just the program sponsor or executives. The stakeholder community can come from the public at large, commuters on a high-traffic road, or even home owners in an area where a new business is being added. The program manager must clearly identify the stakeholders of the program and also ensure that stakeholders clearly understand what the program scope is. Expectations that are not directly aligned with the program deliverables will create a level of frustration and cause a loss of support for the effort. Communicating the program expectations in terms of features, functionality, schedule, cost, and quality must be an ongoing dialogue validating that stakeholders clearly understand what is proposed and how the program will meet needs.

In working with various stakeholder communities, I find that not only must the program managers be able to communicate, they must also be beyond reproach in the information they communicate. Program managers cannot be inconsistent or illogical, or demonstrate a lack of knowledge. They must be aware at all times of the statuses of many projects, efforts, and tasks. They must be able to participate at any level as challenges are brought to them by the project managers on the team, including the progress of the effort, technical challenges, financial issues, and resource concerns. When this pattern is established, the general outcome is one of honesty, reliability, and trust. Honesty and reliability work together to build trust, and it is the trust of the stakeholder community that is crucial to success. A program manager communicating progress who is not trusted is not believable, and therefore the progress, program status, financial estimates, and risk identification all become questionable.

The focus of stakeholder management is to maintain communications and ensure that the stakeholders will benefit from the program, achieve satisfaction, and that the expectations are managed. Figure 4.4 shows

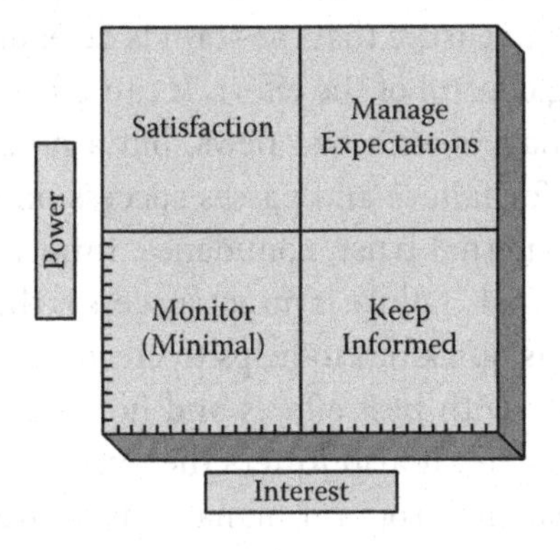

FIGURE 4.4
Leadership study.

the four levels of stakeholder management based on the interest level of the stakeholder and the leverage that it has over the outcome. The lowest quadrant of monitor is used for those stakeholders who will be impacted by the program but do not have direct influence and a minimal interest in the program. Program sponsors will have the highest interest and power over the program and must have expectations managed to ensure that the program meets their expectations and success can be achieved. An executive-level interest in the program may be limited, but success is also measured against the level of satisfaction that is achieved. If an executive determines that the program will not meet expectations or generate a proper return on investment (ROI), the effort may be canceled or modified to meet the desired return.

The reality is that all programs have good days and bad days. There are times when risks are realized, issues are encountered, and the unexpected occurs, causing tremendous concern for the program outcome. Yet a program manager will need to communicate openly and honestly the good and bad of the program regardless of the potential outcry from the stakeholder community. The news must be put into context and balanced between the fears of failure and the realities of issues encountered. A program manager full of all bad news, or all good, soon loses credibility with the community. Therefore the program manager must be courageous in communicating issues by offering potential strategies to overcome the challenges as well as avoiding cheerleading when the effort appears to be firing on all cylinders. In other words, communicate good news quickly and bad news immediately.

At the same time, the program manager is responsible for the work performed by the teams that he or she manages. In the event that an issue or problem occurs, the program manager must assume responsibility for the issues encountered and ensure that the team is defended where necessary, thereby maintaining control of the effort. It can often be painful to stand up to executive managers with bad news, but a program manager who takes responsibility for failure and passes success on to the team creates an environment of internal trust, confidence, innovation, and success. I have always believed that "failure is mine, success is the team's, and failure is not an option." This quick motto helps to create an environment where the team can operate with best efforts and not fear reprisals or attacks in the event of a problem. It also fosters the overall level of communication and ensures that the program manager is aware of the facts, good or bad, and has the ammunition to clearly brief stakeholders. The team that is protected by management becomes more willing to innovate and create, knowing full well that they will not be "punished" for attempting to solve challenges. The safe environment coming out of this leadership style, regardless of the organizational culture, encourages greater internal communication and contributions, eliminating hostile or toxic traits such as finger-pointing, sabotaging, withholding information, and focusing on personal success over organizational.

The traits of courage and expertise lead to a level of confidence that a program manager can achieve with team members and stakeholders. Stakeholders who are concerned about program success will look to the program manager to instill confidence in the objectives. In addition, team members will be more successful following someone leading the team who is confident in the approach. Uncertainty, fear, and anxiety will come across as a lack of belief in the program and potential benefits. This negative will feed the community's concerns, and when issues are encountered, the stakeholder community will immediately interpret its concerns as valid.

On the other hand, a leader who demonstrates confidence in the team, program objectives, potential benefits, and technical approach, and evangelizes the solution will generate a following of stakeholders who begin to believe that success is achievable. It is only through this confidence, communication, and trust that a program manager can ever hope to develop a true HPT. And it is the HPT that can achieve the impossible.

HPTs are achieved through a shared vision, communication, trust, confidence, and motivation. People need to be motivated to move beyond the eight-to-five workday and begin to take pride in the work they are producing. When motivated team members come together to deliver a solution, every aspect of the effort becomes important. An HPT blurs the boundaries between responsibilities as team members begin to help each other and work toward the common goal, setting aside their personal ambitions for a short time and focusing on success. Internal communication increases as the team shares knowledge and issues among themselves, and therefore lessons learned become a dynamic process that takes place throughout program development.

Program Governance

While much of this book has focused on the delivery and motivation of teams, one aspect that is critical is the assurance that critical information required is available and that program and project processes are complied with. Although a transformational leader motivates a team to greater levels of communication, creativity, and innovation, he or she must also make sure that the effort does not stray from the required processes such as reporting on project earned value, schedule performance, risk realization, and status reporting. Skipping over critical process areas of governance because of project momentum is a sure way of creating program failure.

An effective leader will need to ensure that the team complies with all aspects of governance that have been implemented so that the program data, financials, schedule and status are accurate and updated on a consistent basis. This consistently enables program managers to ensure that individual efforts, projects, or tasks do not go astray. A program is a combination of smaller component efforts (projects) and nonproject work that when combined offer a greater benefit than managing each project alone. Therefore, project managers and team members must comply with processes such as schedule management, financial management, performance measurement, and project reporting to support effective program management and provide the foundation by which the effort can eliminate redundancy, disparate efforts, and disconnects between projects.

Although the program manager focuses on leadership compliance with process is one that may require some level of transactional, command-and-control, or authoritative approaches when ensuring that projects consistently and accurately comply with reporting requirements. Obviously, it is better to communicate the end state, vision, and challenges that a program will encounter as well as the necessity and value of program governance, encouraging team members to willingly comply. This can sometimes be a demonstration of courage when standing up to stakeholders as many project managers can be quite skilled leaders in their own right.

Program governance, stakeholder management, and benefit management are critical success factors for a program manager to establish his or her authority over and to contribute to the success of a program. Without effective leadership skills, a program manager can easily alienate groups and decrease the potential for effective communication. The program manager must know the program inside and out and must ensure that all benefits are identified and understood; must communicate when benefits are realized, what the value of those benefits is to the stakeholder community, or overcome concerns if not achieved; must ensure that stakeholders have a chance to address their concerns and be heard on the issues facing the program; and must ensure that projects stay focused on providing the results necessary for the program to achieve its objectives.

Program Life Cycle Management

Programs are managed through a three-part life cycle with program definition, program benefits delivery, and program closure. During this life cycle the program may have a number of projects start and complete, overlapping efforts, transitions to operational control, and the realization of multiple benefits as each project is completed. Ensuring that the program follows the life cycle through the whole process is crucial to success. A program that is not transitioned to operational management at the end is still considered a failure, even when the program benefits are delivered and all projects, initiatives, and operational activities are completed.

When the Denver International Airport was built, the program encompassed building the terminals, putting in place the infrastructure (roads, electrical, water, etc.), implementing a baggage-handling system, staffing the facility, and transitioning the facility over to the operational management team. However, once the airport opened the very first snowfall stranded a large number of passengers at the airport.

The program manager had delivered all of the benefits but did not identify whose responsibility it was to plow the roads from the existing freeway to the new airport terminal. Instead, the city, county, and state all pointed at each other for failing to plow the roads in and out of the airport. Although the program manager completed all of the established tasks and programs, and achieved the benefits established, this minor detail was overlooked, leaving the airport unusable until a resolution was found. It is this aspect of program life cycle management that is crucial to the success of a program. Ensuring that as the program closes all transitional activities are completed and confirming that each operational group has assumed responsibilities for the management of services are as crucial to success as delivering initial benefits or project completion.

5

Distinction between Portfolio, Program, and Project Management

The distinction between portfolio, program, and project management is commonly misunderstood. A program is not a very large project, and multiple projects do not necessarily make a portfolio. Each area has a specific focus and management approach that is leveraged to achieve success.

For the purpose of this discussion, let us start at the bottom (see Figure 5.1) and work our way up. Each level of the chain has varying deliverables and will work with their teams in differing ways. The project manager will have a daily relationship with most of his or her team members and will be working with each to achieve efficient and effective efforts. Project managers often have daily or weekly team meetings but also work on a continuous basis with team members to obtain information, overcome technical challenges, and identify risks for their efforts. The relationship between a project manager and the team is one of support in achieving the objectives of the project and avoiding negative consequences stemming from risks, both known and unknown, that are realized during the effort. In addition, the project manager must have the underlying data on project performance programs comprised of multiple projects and nonproject work for governance and therefore must be constantly updating his or her project plan with actual durations, costs, and resources assigned.

A program manager will manage program-level resources such as program coordinators, PMO interactions, and project managers, but has to step back a bit and place trust in his or her project managers to enable them to perform their job, especially when faced with hardship or dilemma. Project managers are given unique objectives broken down into tasks that achieve a result, service, or product. They are time bound, resource limited, and have very little ambiguity in the effort they are undertaking.

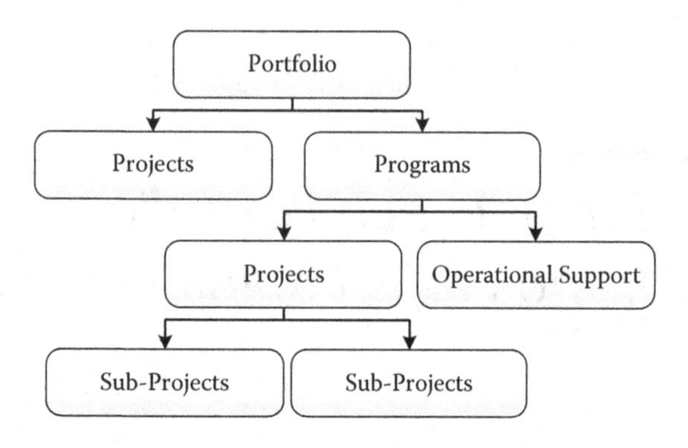

FIGURE 5.1
Portfolios, programs, and projects.

While PMI presents an effective *PMBOK* for project managers, it clearly states that scope is to be constrained and limited whenever possible. Project managers have mechanisms for approving change or scope and minor changes that do not affect the schedule, cost, or risks of the project, but have a Change Control Board (CCB) to approve major changes and, if necessary, a new project baseline from a schedule and cost perspective.

Essentially, a project manager is intended to be myopic, focusing on achieving the deliverables of his or her assigned project and not worrying about the program or organization. However, a program manager is responsible for not only the multiple projects and deliverables along with the success of operational activities but also the strategic alignment of the program and how it will deliver benefits. Program managers need to focus on a much higher-level audience, including the stakeholders from each project as well as project sponsors and programs, while also ensuring that the program remains aligned with the strategic objectives. They must continuously align their programs with the business objectives of the organization and deliver a set of long-term benefits achieved by the project teams. Program managers must focus on the benefits delivered by every project and must clearly communicate the realization of each benefit as it occurs and to the correct audiences. Stakeholders must be informed that benefits have been achieved by managing the program, the program manager must maintain up-to-date schedules and budgets broken down by projects and nonproject efforts. Finally, program managers are responsible for ensuring that the organizational objectives are prioritized and being met throughout the life cycle of the program.

Portfolio managers manage loosely aligned projects and programs. Where program managers are responsible for delivering related projects, programs, and operational activities that together provide a set of benefits, portfolio managers are responsible for managing programs, projects, and other activities in a more loosely defined relationship focused on strategic objectives. A portfolio could be an IT department or a collection of programs and projects for a single client organization. In addition, where projects and programs have a defined start and end date, portfolios are not time bound. For example, an IT department portfolio could be ongoing with programs and projects going through the full life cycle within the overacting the portfolio.

PROGRAM SUCCESS FACTORS

Quite often, program success factors are not identified in advance and leave a gap in the expectations of the program sponsors and the program stakeholders. Success factors define what outcomes will be achieved and what factors will be measured for successful delivery. They are tangible and quantitative so that program managers can manage to the success factors and those success factors can be communicated to team members. Success factors must be identified in advance to ensure that the program manager is well aware of what success will look like, and how to demonstrate that it has been achieved. Many programs have failed to meet all the criteria of the effort simply because success factors were not identified in advance and some expectation was left out of the initiative.

In other words, to be a successful program manager and to build a dynamic and effective team, you must not only have a clear vision of the outcomes of the effort, you must also ensure that you can identify each of the program success factors, how they will be achieved, how to measure the success of each, and finally how to effectively communicate across the program to stakeholders, team members, and sponsors.

Programs can be very successful and meet the goals established, but without a predefined set of measurements and factors that are used to define success, expectations may not be met and therefore the program fails. When success factors in advance are identified, everyone is on the same page and is clear what success looks like.

CASE STUDY: COMMAND AND CONTROL

Throughout this book, examples will be used to help in understanding potential issues that are faced by organizations. Many of these examples will come from personal experience and will be modified to ensure that no identifying information is available. However, please note that each of these cases is an actual real-world example.

> If a rhinoceros were to enter this restaurant now, there is no denying he would have great power here. But I should be the first to rise and assure him that he had no authority whatever.
>
> **— G. K. Chesterton to Alexander Woollcott**

I once had an executive come into a meeting and ignore the challenges, risks, and concerns of the group simply demanding that a time schedule or feature set be included. Experts spoke up and pointed out potential issues, project managers tried to calm the storm by interjecting logic into the conversation, and program participants adapted what was a hardline approach, crossing arms and displaying negative expressions. The result was that the executive became more firm in his or her stance, insisting that this could be done if the team were better qualified or more willing to take on the additional effort. I have even heard some executives refer to teams they have worked with in the past as overcoming similar challenges while facing more extreme situations.

These announcements can come at a huge price to the team. The morale decreases, possibilities of success dry up, risk levels increase, and innovation shuts down. Quite often a command-and-control management style will have executives insist that the problems they face (commitments made to other executives, pressure from customers, or external factors) be the driving force for the project regardless of the realities of the technical or organizational challenges presented by the effort.

1. As a program manager, how can you handle this problem?
2. How do you address the team and the executive to decrease the impact?
3. How can this be avoided in the future?
4. What will be the impact to the program and each component project?

Discussion Questions

The five domains of program management (strategy alignment, program benefits management, stakeholder engagement, program governance, and life cycle management) are critical factors for a program manager to stop and to contribute to the success of a program. These tools enable a program manager to open communication channels with stakeholders, ensure that benefits are identified and agreed upon, and establish control factors necessary for the component projects, providing the opportunity for a consistent set of measures to establish program progress. Without effective leadership skills, a program manager can easily alienate one of the groups and eliminate coordination communication channels. Whether it is with program stakeholders, sponsors, project managers, or team members, a program manager who alienates anyone vital to the program will face an uphill battle to achieving program success.

1. How does a program manager contribute to the success of the program?
2. How can a program manager negatively affect stakeholders?
3. What is the impact of a program manager alienating a team member?
4. What factors define an HPT?
5. How is program vision created and maintained throughout the program life cycle?

Summary

Project, portfolio, and program management are unique management approaches with varying strategies and outcomes utilized for achieving success. Projects are time based, consume resources, and create a unique product, service, or result. A program is a combination of projects and operational activities that together produce a result that a single project could not, and are aligned with the strategic objectives of the organization realizing predefined benefits on a scheduled basis. A portfolio is a collection of programs and projects loosely aligned and tied to an organizational strategy.

The five domains are critical in the ability of the program manager to establish his or her authority and ensure a successful program. Without effective leadership skills, a program manager can easily alienate one of the groups and eliminate many of the success factors for the team. To be a truly effective program manager, you need to focus on clear, concise communications of program vision and strategic objectives so that every team member has a clear understanding of how projects and individual contributions contribute to the organizational objectives of the firm.

The program manager must be able to clearly and concisely identify the program vision and communicate that vision regularly through the life cycle of the program. Vision is something that every team member and stakeholder should contribute to and agree upon so that the final set of benefits is realized and the value of the program is achieved.

In addition, when a leader is capable of building an HPT he or she has created a group that is much more successful than teams that face conflict, distrust, and communication gaps. HPTs work together to resolve conflicts, communicate risks, provide innovative solutions, self-manage, and contribute to program success.

Section II

Leadership

Leaders must be close enough to relate to others, but far enough ahead to motivate them.

—John Maxwell

Control is not leadership; management is not leadership; leadership is leadership. If you seek to lead, invest at least 50% of your time in leading yourself—your own purpose, ethics, principles, motivation, conduct. Invest at least 20% leading those with authority over you and 15% leading your peers.

—Dee Hock
Founder and CEO Emeritus, Visa

6

Introduction to Leadership

> To lead people, walk beside them . . . . As for the best leaders, the people do not notice their existence. The next best, the people honor and praise. The next, the people fear; and the next, the people hate . . . . When the best leader's work is done the people say, "We did it ourselves!"
>
> **—Lao-tzu**

Program managers have a tremendous impact on the culture, the work environment, the project management style, overall team member satisfaction, and success achieving the program objectives. The style of leadership adopted by a program manager can influence the level of communication between teams, individuals, and project managers as well as the manner in which conflict is resolved and the ability of teams to reach the state of being an HPT. Effective program managers demonstrate a leadership style that blends into the culture of the organization but creates a healthy and safe environment for teams to operate in without fear of reprisals, negative conflict, or backstabbing. This healthy environment encourages team members to contribute to success and enables open channels of communication should the project or initiatives not go as planned.

This chapter will identify leadership theories and traits on a broad spectrum that can be leveraged by program managers to enhance their team performance, overcome obstacles, and provide a healthy environment.

Leadership theories are the basis upon which all leaders are measured. Quite often these theories are founded on the examination of great executives and innovators. However, the goal of this book is to focus on leadership as it affects the program and underlying projects, not how to become the next president or CEO of a Fortune 500 company. At the same leadership tools a CEO or entrepreneur uses can also be tailored to program management and building HPTs.

Every team, culture, and environment differ in their expectations of a leader and their needs for the role that a program manager will play. In some organizations, they have been trained to defer to the management structure and stay quiet about their concerns, following orders regardless of the outcome; whereas in others they challenge the leader in an attempt to work together creating positive conflicts that generate greater innovation and productivity. Regardless of the team or the organization, my style often moves between leadership styles to bring teams together based on the environment and program needs. While some of these styles lie outside my comfort zone, they are still useful in bringing teams together into a cohesive and collaborative group.

During the next few pages, we will delve into leadership research at a deeper level both to understand the history of leadership and to identify the various styles that are most recognized. The goal here is to ensure (1) that when we discuss leadership, we are doing so from the same page; and (2) that we have a clear idea of how leadership can create an HPT that outperforms expectations; and (3) how one style may result in negative interaction whereas another style may create an environment where everyone gets along but no work is actually accomplished. Although much of this book has been written based on real-world situations and experience, this section describes a bit more research that has been performed by various authors on the subject of leadership. Much of this part is paraphrased, but whenever possible I have referenced those researchers, and all references are available in the back of the book. I encourage you to read up further on some of these authors so that you can better form your own opinions of the research studies.

As you read through the next chapter, consider your own leadership style and the effect that it may have on the teams that you lead, how you could have adopted a differing style, and whether that would have had a positive or negative influence on the team. In other words, although you may feel a certain style is comfortable and the most effective, consider the various styles available and ask yourself if there were better ways to work with your teams. Was there a single approach you could have used? A hybrid approach? Was it a single leadership approach that made your effort a success or was it a combination of styles ranging from one side of the spectrum to the other? Did you try different styles of leadership to see what was effective or did the needs of the team force you to lead from a particular perspective? Consider the environments that you have been in and your willingness to morph to those environments regardless of the positive or negative influences they had.

LEADERSHIP VERSUS MANAGEMENT

> People ask the difference between a leader and a boss. The leader leads, and the boss drives.
>
> **—Theodore Roosevelt**

With the extensive literature on leadership, discussions on the subject should start with a clear description or definition of what effective leadership is. Topping (2001) offers that when evaluating leadership, it should be recognized that not all individuals are capable of becoming true leaders. As a matter of fact, most will not advance beyond the single skill of being a manager rather than a leader simply because of training, education, empathy, or personality-related issues. The whole package of a leader requires a much more comprehensive model that can leverage inherent abilities in the individual, while utilizing learning and history to improve upon process. Every month there is a new set of books on leadership and various approaches. So why do we continue to buy these books? If so many smart people have been studying leadership for all these years, why don't we have the answers? The answer, of course, is that there is no answer—at least no one right answer" (Topping 2001, 3).

Most books on leadership have some value that can be useful to a particular individual in his or her professional style. The driver of these books is that leadership ability can be learned and is not inherently found in a minority of individuals.

To date, there are more than 150 definitions of leadership, but one definition of leadership is relatively standard: "Leadership is the reciprocal process of mobilizing, by persons with certain motives and values, various economic, political, and other resources, in a context of competition and comfort, in order to realize goals independently or mutually held by both leaders and followers" (Burns 1978, 425).

Leaders are described as individuals who are intrinsically motivated, self-managed, demonstrate excellent communication skills, and are visionary, empathetic, and naturally charismatic. A leader is someone others choose to follow and to support and someone who can get others to set their personal objectives aside to pursue a new goal contributing to a more common objective (Hogan, Curphy, and Hogan 1994). Leaders don't have to be executives or have a title that grants them authority; rather they are people who have gained the skills, whether inherent or learned, that motivate people toward a common goal driving innovation and excellence.

An effective leader pools individual team members into a more comprehensive team that when working together can achieve more than individual efforts would be able to. Leaders don't manage or mandate actions or tasks; instead they motivate and empower staff to identify and complete the work necessary for the established outcome. Teams led through effective leadership minimize risk, transition conflict from negative and unhealthy to positive and innovative, and as a general rule are capable of exceeding expectations through joint efforts, open communications, and clear lines of responsibilities. It is through leadership that a vision can be established to ensure that team members understand the outcome, product, or result that the team is trying to produce (see Figure 6.1). This vision is a clear, concise statement, easily understood and repeated often so that individuals, stakeholders, and teams can work toward a common objective avoiding the consequences of ambiguity and confusion.

Leadership is the motivating and building of teams to achieve an established outcome by creating a positive and healthy environment and utilizing communication channels, conflict resolution strategies, team

FIGURE 6.1
Leadership traits.

building, and development of clear roles and responsibilities so that teams more effectively work together. Therefore the result of leadership is the positive achievement of initiatives in the pursuit of goals and objectives that are beneficial to the organization's strategic benefits and have a positive impact on stakeholders. The focus on motivation, comfort, and reciprocal process all imply that the leader is working with the team rather than directing them.

One of the more common misunderstandings around leadership is that it is management. It is not. Management is instructing personnel, timekeeping, governance, or mandating directives and tasks. Management is short term, focuses on the bottom line, and does not work with staff with an eye toward motivation or empowerment. On the other hand, regardless of the leadership that we have in an organization, management is also necessary and will never be replaced. Implementing and directing administrative actions, focus on the bottom line, and short-term visions are necessary for the achievement of long-term objectives.

Ineffective managers build a hostile and troubled environment where conflict is not resolved and blame is often heaped on individuals. Teams working with ineffective managers are often risk intolerant, unengaged, and avoid innovation rather than drive toward success. Negative environments such as this are rarely addressed because each program is different, and reasons for program failure are commonly spread throughout the team. We often find that although team members predict failure early in the process and they may share that opinion internally, most don't communicate their concerns directly to management.

Whereas management focuses on handling complexity, leadership is centered on change and innovation in the organizational environment. Leaders drive the innovation, motivate teams, empower staff, and open up lines of communication; managers, on the other hand, deliver on the tasks that have been assigned. Leaders are often recognized as having most, if not all, of the following traits:

- Charismatic
- Transformational
- Visionary
- Trustworthy
- Courageous
- Confident
- Motivational

- Innovative
- Possessing effective communication skills
- Driving the empowerment of staff

Discussions on leadership should start with a description or definition of what effective leadership is: "Leaders may be born or made or a combination of the two theories . . . but leadership is the ability to turn vision into reality" (Ilan and Higgins 2005, 28). Leaders generally motivate followers by generating goodwill, confidence, and security. Benefits are longer term but are focused on the program's success rather than on its failures. Leaders drive positive behavior through motivation instead of directing or commanding it.

As such, leadership must be able to achieve objectives by increasing the value of a group over the value of an individual: "Leadership involves persuading other people to set aside for a period of time their individual concerns and to pursue a common goal that is important for the responsibilities and welfare of a group" (Hogan, Curphy, and Hogan 1994, 279). It is important to note that leaders can be found at all levels of an organization, and title, position, or management role do not play into the definition of leadership.

Leadership is focused on persuading others to contribute to a common goal or a common set of objectives, moving beyond personal goals and contributing to the defined team goal or vision. It can range from driving an organization or large corporate entity to leading a small team of key players. The ability of a leader to achieve common goals by inspiring others to follow him or her to a common objective implies that the impact of a leader is not the actual production, innovation, or output but instead the output of the team as it follows the leader: "Leadership concerns building cohesive and goal-oriented teams; there is a causal and definitional link between leadership and team performance" (Hogan, Curphy, and Hogan 1994, 282). The output of the team is the basis upon which the quality of the leader is measured.

Leaders are definitely not managers, although leaders can often take on the role of managers just as managers can also hold leadership traits. Managers focus on the delivery of services according to a predefined set of specifications. Although a manager could also have leadership ability, we look to managers to deliver consistent results based on the needs of the organization.

Management focuses on handling complexity, whereas leadership is centered on change and innovation in the organizational environment (Kotter

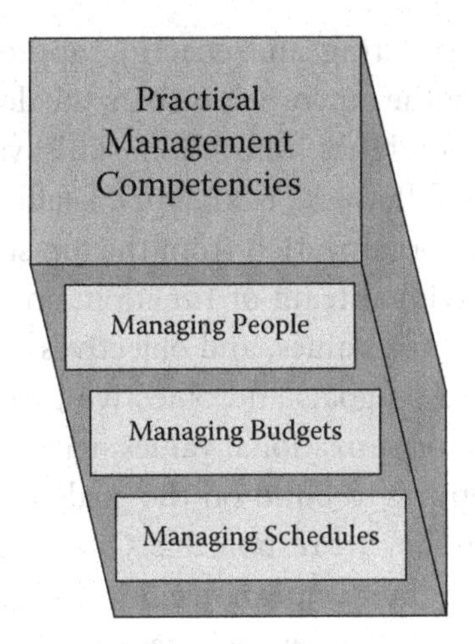

FIGURE 6.2
Traditional management competencies.

1980). Though leadership can complement management, it will never replace it. The manager has a short-term vision looking at administrative actions, focusing on the bottom line, and doing things right, whereas the leader looks long term, innovates, and does the right thing (see Figure 6.2). The leader directs the activities of a group to achieve a common goal. This common goal must be not only understandable and agreed upon but also communicated to the team in such a way as to ensure that the team clearly understands the objective and willingly follows the goal.

As shown in Figure 6.3, an important distinction here is that a manager focuses on doing things the correct way according to the predefined system and procedures while focusing on the short-term deliverables assigned to him or her. A leader is an innovator who generates ideation, drives change in an organization because of authority granted to the individual, and enables people to follow him or her because of a common and shared vision. Whereas a manager must be followed in an organization, a leader is someone who people choose to follow based on the vision, passion, communication, and knowledge of the leader.

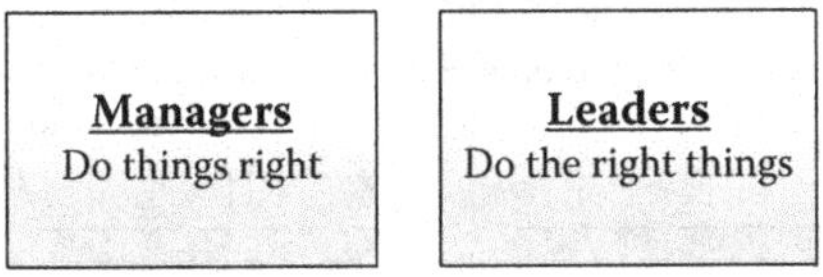

Managers	**Leaders**
Do things right	Do the right things

FIGURE 6.3
Management versus leadership.

A leader's skills in selecting and enacting appropriate behaviors are related to the success of the outcome. As such, the definition of leadership is a complex one that combines innovation, skills, values, ethics, change, results, and support of followers (Yukl 2002). Leadership can be found at any and all levels of an organization from the top strategy-setting executives down to the individual team or functional manager. Those leaders who align with the culture, values, and objectives of a firm can help that firm in pushing toward its goals and objectives, while negatively driven leaders who clash with organizational values may push against the objectives instead of attempting to build on the goals for the organization or business unit for which they are responsible. One of the key traits of leaders is neither their title nor their importance in the organization but instead the willingness of others to follow, their desire to lead, and a clearly communicated vision along with a set of ethical values (see Figure 6.4).

The value of leadership is something that can be difficult to measure. Determining who is a leader will not be accomplished through a simple test that outlines a leader and what his or her style is. Because of situational factors, leaders who are effective in one environment may not be so in another. Leadership is generally best measured by the perceptions and accomplishments of the team, something that cannot be forced but instead is accomplished through the willing participation of the followers: "Leadership only occurs when others willingly adopt, for a period of time, the goals of a group as their own" (Hogan, Curphy, and Hogan 1994, 282). More importantly, an effective team functions together in a manner that is noticeably different from an individualistic organization: "There is a palpable feeling of excitement that arises when you work with smart

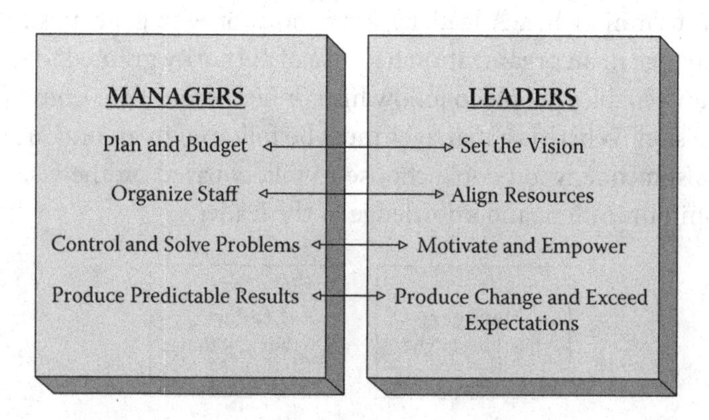

FIGURE 6.4
Management and leadership functions.

and engaged people whose goals and commitment are the same as yours" (Knight and Dyer 2005, 12).

Because leadership is challenging followers to a new, clearly communicated, and advantageous vision, and many organizations have cultural or environmental barriers, the need for challenging the organization's existing environment and introducing innovation against the change resisters of the firm will require that leaders be willing to accept that their positions are not guaranteed and therefore that they will need to go against what is fundamentally a psychological barrier to risk taking. Leaders must be able to take risks for success and leverage the underlying organization core competence. The risk faced by those in leadership roles can result in project, program, divisional, or corporate failure, and in turn may result in job, reputation, or professional standing loss. Therefore leaders must believe in themselves and that goals are achievable and realistic.

Although program managers cannot necessarily change the organization as a whole, the innovations in a program can invoke organizational change, increasing the firm's ability to deliver products, services, or results. Quite often the approaches taken by a program team will help a firm to set new best practices and provide adoption of best practices by other leaders. While the program achieves success, the organization learns and can adapt approaches to enhance other functional areas. In firms that have executive-level leaders who are working to enhance their organization, they will recognize how the approaches taken by the program will assist in other functions and develop change strategies for the organization as a whole, pointing to the program as a successful example.

All programs must be clearly aligned with the organizational objectives so that the innovational approach also aligns with the goals of the firm. If leaders drive change in a direction that is contrary to the organizational strategy, they can hinder the firm's ability to achieve its targets and instead create conflict within organizational elements. A leader whose belief set is not in line with the organizational focus and strategy can be "ruinous" to achieving strategic objectives. In program management, we are often faced with taking on efforts that will push an organization toward a new direction or leverage new technology. As leaders, we have to ensure that the approach is supportable by the firm in the long run and will not create problems when transitioning to functional and operational departments. In other words, a program that can't be supported by the environment may achieve success in the delivery of benefits but would fail in the long-term operational perspective.

To be successful, an organization must be able to challenge and effectively change the status quo and introduce change as a disrupter to the standard processes and belief sets. The disruption to status quo changes the organization and creates innovation. In this way, program leadership can be risky by challenging what has always been done to define a new set of approaches, thoughts, and ideas. Program managers will often struggle against those who are comfortable with repetition to develop new approaches that challenge the existing processes, as well as to bring some level of creativity and introduce new thoughts into the mix. They will frequently attempt to invoke change against which others are resistant and must be able to leverage the organizational objectives, balancing between the cooperation and conflict of team members, in an attempt to meet current needs, while still extending change toward future prospects to meet the needs of the innovation.

Innovation drives an organization further toward its defined goals, productivity, and effective use of resources through the implementation of change. Self-motivated growth within an organization leveraging the resources available is called "internal entrepreneurship." As such, Morris and Kuratko (2002) identify several obstacles to internal entrepreneurship that drives internal change:

1. Rigid and oppressively controlling corporate systems and bureaucracies;
2. Multiple-level hierarchical structure with top-down management, restricted spans of control, and restricted communication channels;
3. Myopic leadership vision and lack of commitment for entrepreneurial activity by senior leadership;
4. Highly structured policies and procedures with long, complex approval cycles and documentation requirements;
5. A cautious, short-term organizational culture that instills fear of failure and resistance to change. (12)

Just about everyone I meet has a story about a really ineffective manager they have worked with. As a matter of fact, most of us have had the dubious *pleasure* of working with someone who is narcissistic, egotistical, or uninformed, or more politely, someone who was less than competent, insisting on wasting valuable project time on meaningless questions and discussions. These ineffective leaders are often ridiculed behind their backs and find very little support from the team. While not respected, or trusted, they often become more of a problem than a solution.

Not too long ago, I walked into a program based in the software development division of an IT department where the chief product manager was eventually fired for failing consistently on programs and projects. Although this leader felt that he knew project management, he was neither versed in project management methodology nor leadership, and his software development knowledge was outdated. When failure was looming, he would heap insults, accusations, and blame on the team members involved. He would consistently report that while he told the team what to do, they just didn't get it right. Unfortunately, because of his position as an officer of the firm, only a few people complained to the company about his bad management style and his abusive and derogatory comments to employees. I mention this only because it created an overall hostile environment where derogatory comments and personal attacks were acceptable and common. These negative approaches were inflammatory to the staff and increased the hostility level, further decreasing morale and leading to reduced communication and distrust among team members.

While oftentimes looking back at bad management can be humorous, it begs a number of questions about the impact that the bad manager can really have on an organization. Empirically, individuals in those scenarios can evaluate their own actions and the general results of peers, but without evidence it remains opinion. Therefore understanding the impact of management is vital to understanding what most workers inherently realize: bad managers kill organizations. Understanding the actual impact that a bad manager can have on the staff is of great value to all of us.

7

External Factors Affecting Leadership

> Great leaders are almost always great simplifiers, who can cut through argument, debate, and doubt, to offer a solution everybody can understand.
>
> **—General Colin Powell**

While this book is focused on the value leadership skills can provide to program management and building high performing teams, it is important to recognize that a program manager will be pulled in many directions and affected by a long list of external influences in his or her day-to-day management. The focus is, of course, on delivering the benefits of the program, achieving cost and quality standards and managing the efforts that are necessary for achieving success, yet a program manager in a real world setting will face a number of challenges coming from the organization itself, the stakeholder community, supporting departments, as well as program sponsors and in some cases the program manager's direct supervisor.

STAKEHOLDER COMMUNITY

Regardless of the group or individual, a program manager will spend a great deal of time and effort working with external factors taking him or her away from managing the program itself. As mentioned in previous chapters, the creation of a high performing team requires that the team feel comfortable communicating, innovating and working with positive conflict resolution strategies. The program and project managers of an effort create this environment and to be truly effective, run interference for any external influences that can have a negative impact on the team environment.

Programs have benefits that once realized will affect a community of stakeholders. While a program is made up of projects and non-project related work, each effort has a specific set of stakeholders affected. Not all projects will have the same individuals in their stakeholder community though there very well might be overlap. However, the program manager is responsible for all stakeholders affected by the program itself. Communication with all stakeholders with a clear, concise message and ensuring that there are open channels for communicated is a key success factor for any program. Stakeholders must be aware, and often reminded, of the benefits the program will deliver as well as potential issues that may arise as the effort progresses. They are also a great source for identifying potential risks and ensuring that expectations are clearly set avoiding issues down the road.

Each project manager should have a communication plan for dealing with the stakeholder community, and the program manager may often participate in that plan. Directly communicating from the program perspective ensures that the overall message is communicated and stakeholders are not lost in the project specific deliverables. These communiques can take the form of reports, memos, emails, or meetings and will provide the program manager with valuable feedback on the defined benefits as well as potential hurdles that may be encountered. However, in these interactions, scope creep and enhancement requests will generally be consistently asked for. In addition, complaints and issues encountered by the stakeholders will be communicated along with any negative concerns as to the potential program success or failure.

The program manager is responsible for maintaining the open channels of communication but also protecting the team from interference wherever possible. While scope creep is a realistic expectation, every request will not be approved for the program and quite often debates will take place as to cost versus benefit for new asks. An effective program manager will work with these stakeholders and their project managers to handle all of the external influence until the point that the issues or requests are approved for the program. The successful manager will keep their team in the loop on these discussions but will not allow the external negotiations and communications to detract from the team's effort. Instead the team needs to be protected to a point where their focus can be maintained and their progress uninterrupted. Negatives from either the stakeholder community or the organization itself should not be allowed to affect the team morale or confidence.

If issues identified need to be incorporated into the program or project efforts, the manager will bring these changes to the team leveraging the open communication channels and promoting positive conflict based discussions to drive discussions and identify potential risks or benefits that can be achieved.

ORGANIZATIONAL CULTURE AND PROCESSES

In addition to the stakeholder community, there are also a tremendous number of other external factors that a program manager will work with and often run interference for the team. Organizational culture and processes can have a huge impact on a program. Once a program team is set up, they need to be disconnected from any negative cultural aspects, maintain the positive ones and given a chance to grow into a effective high performing team.

Leadership builds a team but organizational factors can tear it down much quicker than the time it took to build. Quite often program managers will work toward collocating teams, creating a bubble for the team to work within the organization and facilitate an environment of positive and open relationships for team members. If the organization, or client groups, present a less than positive environment the program manager will work to keep those negative factors away from the team ensuring that successful team development and management can be promoted.

Therefore, while a team needs open communication with themselves and to understand the stakeholder needs and requirements, the program manager is there to influence those interactions maintaining a positive environment and decreasing the potential for a negative or hostile environment.

Furthermore, the most effective programs and projects follow a methodology such as the PMI Program and Project standards. While an organization may have internal processes that have to be taken account of in the program, a successful program manager will integrate external processes into their established program and project methodology to meet external needs but allow the team to focus on a well proven approach to managing a project.

For example, the procurement process may require executive sign-off on purchases greater than $50,000. In a case such as this, the program manager will tailor the project management procurement process to elevate certain requests so that they can directly intervene and work with the organization to achieve the necessary approvals and ensure the project is not delayed by the external directives.

External processes cannot be ignored and must be planned for. A successful program manager will research these factors, modify the program and project management processes to incorporate the external factors, and assign roles to ensure that the external requirement are met, oftentimes by a program manager rather than a team member.

Internal/External Auditing

Quite often various organizations will have external auditing processes scheduled. A financial firm may have an audit scheduled yearly or quarterly while an IT organization that was certified with something like ISO would have annual audits and a 3 year recertification cycle. Program managers should research these factors in advance; ensure that these scheduled (and sometimes unscheduled) factors are included in the planning process. Governance can then be used to ensure that all records are constantly updated and in a form that is ready for audit avoiding having to do rework or drop project efforts to update information.

If the project and program management processes are modified to ensure records are maintained in a real-time basis and the program governance is assigned the responsibility of verifying the records, the project team can focus on the tasks at hand without interruption or disturbance.

Legal/Regulatory Changes

In many organizations, legal and regulatory issues drive the business itself. Healthcare is updated yearly based on the information required for Medicare and Medicaid reporting purposes. A program that spans multiple years with projects focused on achieving a set of defined benefits may have to add additional project resources that are not specifically assigned for program benefits but instead are ready to respond to the potential impact of legal or regulatory changes. In addition, a SME may be assigned to monitor the regulatory agency ensuring that any upcoming changes will not adversely affect a project or the established program benefits.

Program managers in the planning stages will research both the organization as well as the external factors to ensure a clear understanding of how external factors can affect the effort. From this research plans can be established to avoid interruption of critical work efforts and can be included in the program/project planning process as well as finance estimating ensuring that productivity, schedules and budgets are not adversely affected.

Executive Leadership

Executive leaders have a tremendous amount of external factors and priorities affecting their decision making. Priorities shift, competition introduces new products, new innovations are instantiated, client complaints and issues become critical and financial concerns are always a major factor.

Programs that are instantiated with long term benefit realization plans will be affected by the executive leadership team and quite often the benefits of the program may not be as critical as a new concern. The program manager is responsible for constantly monitoring this concern, maintaining an up-to-date ROI as well as benefit realization plans, program spend rates and projections to completion. To be effective and to ensure ongoing support from the organization, a program manager will run interference for the effort and ensure executives are in the loop and reminded of the benefits that the program will occur. In some cases, reverting to ROI and demonstrating spend to date rates can assist an executive in understanding that interrupting the program in lieu of a new priority may be more costly than seeing the program to completion.

Finally, projections to completion can demonstrate the amount of money for the program to complete if it were left in place. Oftentimes, leaving a program to complete can be more cost effective than interrupting the effort and can help executives understand the impact of interrupting the effort midstream. While other priorities may become critical, this information is extremely valuable for executives to better understand the cost to impacting the program. From this knowledge base, executives can make better decisions and truly understand the impact of canceling or stopping the program.

While this may not cause the executive leadership to maintain the effort in the middle of the crisis, the program manager will have done an effective job of communicating value, costs, and impacts for closing or interrupting the program and the executive leadership team will be better informed in their decision making.

Geographically Diverse Teams

Often, outsourcing is a cost effective way for a organization to procure resources with valuable talents at an inexpensive cost for short periods of time enabling the use of SME's to assist in achieving goals. As such, program managers have a distinct need to be well versed in working with outsourced and often geographically diverse teams. These groups can be located throughout the country or the world and may bring with them culturally diverse values, ethics, and experiences. In addition, language barriers, time differences and technology challenges can often cause issues with building teams.

Program managers must first understand the contracts between these outsourced vendors, service level agreements, performance metric standards, and responsibility levels. From that point, the Program manager needs to embrace these external vendors as part of the core HPT bringing them into every meeting, dialing them up for impromptu discussions, and sharing documentation, requirements and goals of the effort. They are brought into the team building experience, assigned roles and responsibilities and are held to the same standards as the internal team members.

A successful program manager will embrace these outsourced teams as part of the HPT and will not allow for completion, alienation, or distance to impact their critical role in the success of the program. Quite often I have traveled to outsourced team sites to meet them individually and have brought them to our internal sites to meet their team peers. Through this, the outsourced team has a chance to establish relationships, communication channels and to feel that they are part of the team in achieving the successful delivery of program benefits. Any technique that avoids outsourced team members from feeling as if they are outsiders or internal team members as seeing the outsourced teams as competition are valuable in achieving a truly high performing team and the program benefits.

8

Individual Motivation

> All of the great leaders have had one characteristic in common; it was the willingness to confront unequivocally the major anxiety of their people in their time. This, and not much else, is the essence of leadership.
>
> **—John Kenneth Galbraith**

Psychoanalysis is used to open a window into individual behaviors, as well as into organizational environments, cultures, and management theory. It promotes a better understanding of how environments and management interactions can affect an individual by directly analyzing the impact on the unconscious mind and in return the impact that the unconscious mind can have on individual behaviors. Psychoanalysis confronts the unconscious mind in a formalized manner evaluating the implicit and explicit impacts that culture, environment, and leadership have on it (Foucault 1970).

Organizations form a unique culture and environment that can range from a positive or negative perspective. They demonstrate, through the interaction of individuals, a society where social dynamics are acted out. Each will vary in the manner through which culture and social norms are established and demonstrated, but cultural trends such as authoritarianism or narcissism weave themselves into the psycho-structures of organizations, affecting leadership, communication, and group relations (Carr 1993; Lasch 1980). No organization is either 100 percent positive or negative; it exists on a spectrum where there are many shades of gray. The psychoanalytical approach focuses not on the right techniques to motivate, but instead on the concept that through work people pursue many different conscious and unconscious aims, and that organizational culture has a tremendous impact on behavior (Gabriel and Carr 2002).

Though organizations create innovation, creativity, and have unique levels of risk tolerance, they can also create an environment that breeds anxiety if goals or objectives are not met. Each organization holds its staff responsible for individual performance, oftentimes forcing people to work together in performance tasks that are not enjoyed, and can be found treating employees in an impersonal and cold manner while requiring those same individuals to demonstrate loyalty and support to the organization itself. The agreement between a company and its staff is that individuals are paid for their work effort, yet negative impacts can occur in performing the work. Negative impacts can create feelings of fear or anxiety which, in a work environment, can be seen as an "incapacitating emotion which individuals defend themselves against through the mechanisms of defense" (Gabriel and Carr 2002, 35).

In "The Ego and Id," Freud (1923) evaluated individuals and defined motivators that explain the need to achieve gains such as satisfaction in creation or solving complex problems, personal rewards, or maintaining a level of safety and security. Individuals have their own conscious and unconscious motivators driving their performance and investment in their jobs. Whether those are growth, job stability, increased income, or ego driven, the reasons that people perform in their jobs can be complex.

The negative impact of anxiety resulting through organizational interaction is that individuals "often resort to dysfunctional routines which stunt creativity, block the expression of emotion or conflict, and above all, undermine the organization's rational and effective functioning" (Gabriel and Carr, 2002, conflict, and above all, undermine the organization's rational and effective functioning' (Gabriel and Carr 2002, 356). These defenses against anxiety can create collective delusions causing a fight-or-flight reaction from nonexistent threats while ignoring real issues related to the work at hand. In addition, the defense mechanism in most individuals responding to anxiety or implied attacks is a negative one, diminishing innovation, risk tolerance, and support for the efforts at hand; "excessive anxiety leads to highly dysfunctional defensive routines, while inadequate anxiety breeds complacency, inertia, and gradual decay" (Gabriel and Carr 2002, 369). When a level of anxiety is maintained for a long duration of time, individuals begin to lose interest in their current positions and attempt to modify their professional situations through job searching, disengaging, and decreased motivation, leading to low morale and the inevitable outcome of increased attrition.

Leadership is instrumental turning negative behavior patterns to positive ones and creating environments where individuals value the challenges they face. Understanding how the organizational culture has affected and trained its staff is crucial to understanding the type of leadership approach that would have the most impact. A staff that operates with high anxiety and negative consequences will be hesitant to offer suggestions, engage in open communication, offer innovative concepts, and identify program risks. They will maintain the status quo, following the same path each time to avoid personal attacks and blame for failure even when failure is anticipated. A leader must be able to recognize this pattern when building the team and create a safe environment for the team that enables them to feel comfortable within the program, and encourages openness and innovation.

This includes positive conflict resolution, public acknowledgement of innovation, and consistently running interface from outside elements so the team can build trust in a comfort zone. The program manager will need to protect team, allowing them to operate in a "bubble" within the organization and to begin to function without fear of reprisals or negative consequences. The program manager must step forward to protect his or her team and facilitate conflict resolution and internal communication by building a safe and healthy environment that can be sustained for the program's duration.

Studies have shown that various "interventions" can help to shape the impact of organizational health and could more positively impact the employee reactions. As a program manager, your goal is not to change the organization but to understand the culture and to create an environment for your team that is safe and trusting and increases communication channels among team members. A program manager does not focus on the organizational issues; instead the goal is to operate within it, creating a safe and healthy environment in which your team can function. To achieve this, a program manager must understand the individuals' motivators and what drives them to success, while also determining how the organization health affects them, especially when making difficult decisions, facing constant crises and ethical issues.

The organizational influence on team members may affect each member in a different way. One may be resentful of the leaders he or she has worked for in the past, and another may have had very positive experiences. Each member needs to be treated as an individual, and management styles must be adjusted to leverage the most successful approach possible. If you have an individual who is a resister to change, you may need to take a more authoritative approach, pushing the team member toward a task, while others may just need to see you display confidence in their work and reward successes as positive.

There are a number of ways in which individuals, regardless of their position in the organization, are motivated. Some are consciously and purposely driven by organizational management, such as rewards, incentives, job sharing, and performance measurements, while others fall into the more subtle categories of unconscious drivers, related to both personality and needs—things like intrinsic drive, innovative nature, and self-management.

It was actually Hewett, not Freud, who uncovered the initial outlay of the unconscious. In his book Elements of Psychology (1889), Hewett declared that "unconscious knowing and unconscious willing are phases which defy all interpretation" (32). Freud later added to this by pointing to the unconscious as a motivator and the conscious mind's way of "hiding" thoughts and desires from awareness.

Gabriel and Carr (2002) found that much of the activity responsible for behavior is hidden from the conscious mind and driven by unconscious thoughts, drivers, and motivators. This is in turn demonstrates that the unconscious is driving actions and using tools such as anxiety to trigger defense mechanisms. These unconscious drivers often have a dramatic influence over the individual's performance. If the organization is more toxic with negative interactions, a team member may fear for his or her job and the security that the position brings. In such a situation, the team member may unconsciously react in a more risk-adverse manner, staying in the background and maintaining the status quo. A program manager will need to recognize the individual and his or her drivers to encourage change and empower the staff to innovate and accept risk as part of the program process.

Pienaar (2008) identifies three characteristics leading to leadership ineffectiveness: "character of a leader, the ability to manage one's own emotion, and difficulty in effectively managing interpersonal relationships" (133). It is the ability to manage one's own emotions that can be most affected by Maslow's hierarchy of needs, because a leader who separates emotional reactions from intellectual is a successful one. The subconscious drivers that attempt to achieve stability in safety and physiological security can dramatically affect an individual's emotional state. Self-deception can be added to these characteristics, leading to ineffectiveness as both a team member and a leader. Self-deception is a trait that causes leaders to miss out on their own weaknesses and improperly inform themselves as to what their strengths may be. Furthermore, the unconscious drivers that create defense mechanisms, such as anxiety and fear, from an individual's needs set can even be hidden from the individual's conscious mind. Successful program managers self-analyze their reactions and interactions with teams to better understand why they have reacted in a particular way to various situations. Program managers must understand themselves and work toward a positive leadership style instead of allowing unconscious drivers to affect their own or their team's behavior.

Unconscious thoughts, feelings, and experiences demonstrate themselves through defense mechanisms such as "regression, rationalization, denial, sublimation, identification, projection, displacement and reaction formation" (Waelder 1967, 33). These defense mechanisms may reduce the conscious mind's stress and anxiety, but the unconscious mind retains the experience and will influence behavior without active awareness or conscious thought when faced with similar situations or environmental factors.

The resistance of the unconscious mind to identify or explain it creates a challenge to identify what factors are influencing behavior beyond the consciousness. With so much time spent within organizations, the impact of the organization on the unconscious is tremendous. Gabriel and Carr (2002) propose that all people's "dreams, anxieties, fears, impulses, emotions, and fantasies are rooted in their experiences as members of organizations" (352).

In 1943, Maslow introduced "A Theory of Human Motivation." This article provided even greater insight into the human motivators and drivers, especially as they relate to the unconscious mind. Maslow describes human needs in five distinct phases:

1. Physiological
2. Safety
3. Love/belonging
4. Esteem
5. Self-actualization

These needs are hierarchical based on requiring a person to achieve a level of confidence in the need before moving on to the next level. Any disruption in needs at a particular level will force the person to regress to earlier needs. So each level must be satisfied in turn in order for the individual to begin pursuit of the next.

Physiological needs relate to satisfying basic living, such as breathing, food, water, sex, sleep, homeostasis, and excretion (Maslow 1943). If any of these are not met, the individual will find him or herself with a greater and greater "hunger" to satisfy them. If the needs are consistently not met, the person will be consumed with the idea of satisfying them to the exclusion of higher needs. Therefore, if employees are constantly threatened with the loss of a job or the ability to make money, they will begin to focus their efforts on maintaining their position regardless of what it takes. Ethics, morals, and values can often be impacted when employees are fighting for their position within a firm.

Once the needs for safety have been realized, a person can move on to the next level, love, and social belonging. Focused on friendship, family, and sexual intimacy (Maslow 1943), the individual who has achieved this level has a general comfort established with physiological and safety needs, and has the additional time and energy to pursue social belonging. It is at this point that team members can afford to build friendships and create trust among coworkers. To establish team development, employees must at least reach this level or they are constantly in competition with their peers, trying to protect their position and employment status. If at any point in time any of the lower-level needs are threatened or not satisfied for a period of time, the "hunger" to satisfy those needs again becomes a predominant driver.

Esteem starts to move the individual much closer to self-actualization. Esteem focuses on self-esteem, confidence, achievement, respect for others, and respect by others (Maslow 1943). It is at this level that individuals can start to enjoy the satisfaction of lower-level needs and begin focusing on areas of greater value to their psyche. Here employees will begin to expand their boundaries and begin to look outside of themselves for more information and knowledge, and to leverage the experience of others. Employees who reach this state can afford to be innovative, invest time and energy in learning what approaches others have found, and research problems.

Finally, Maslow points to the highest level of his hierarchy, self-actualization. It is at this point that an individual who has satisfied lower-level needs will be able to focus on the more cerebral aspects of his or her life. Self-actualization looks toward satisfying needs such as morality, creativity, spontaneity, problem solving, lack of prejudice, and acceptance of facts

(Maslow 1943). This is the level that we look to for leaders. Leaders are confident in their abilities, not threatened by the environment or culture, and able to create vision and approaches for programs to achieve the benefits that are demanded.

Team members who are worried about safety in terms of job stability may revert in behavior to trying to ensure that their job is safe before focusing on creativity and risky approaches. This reversion will leave them unable to grow with the team and to take on the challenges presented by the program.

No matter what level the individual has achieved, the failure to satisfy lower-level needs will result in their facing anxiety, fear, and other unconscious drivers until these lower-level needs are satisfied. The goal of a program manager is to understand where each team member resides on the scale and to work with them to achieve a level where their best skills and efforts can be brought to bear on the program. Without understanding this, employees are on their own, struggling with concerns about their position within the organization or potential job loss if the program is not successful.

Determining what impact organizational environment can have on an individual, especially when that person's safety is threatened, is a crucial aspect. It is the negative organizational environment that can cause great duress to team members. Neurotic or toxic organizations that have a negative environment capable of impacting individuals and potentially exposing these individuals to moral dilemmas are most likely to require a shift in leadership style to protect the team and create a positive and creative environment. Most importantly, leaders operating in a hostile or toxic environment must rely on their skill sets, leadership styles, and communication skills to avoid being dragged into the negative culture of the organization they operate within.

I once worked in an environment where the manager of the department would actually look at his employees in a meeting and ask, "Why do I pay you for being such an idiot?" This environment was demeaning, negative, and demoralizing. While we operated as consultants to move a program forward, one of my staff was called in to the manager's office and berated for over an hour regarding something she had not done. When she left, she immediately came to me and thanked me for her job with my firm. Her stance was that the manager could kick her off the project, but only I could fire her; she knew she was safe with me enabling her to operate within the negative culture while also keeping her ability to do what was right and necessary to move the program forward without concern about losing her job or allowing the organization to demoralize her. It is a prime example of program managers creating an environment of positive culture even within an organization that is culturally challenged.

As has been discussed, leaders come in many forms and styles. No single trait can be identified to ensure effective leadership, and the morality of the leader has a tremendous impact on the organization, followers, and effectiveness of the leader. Therefore an examination of the individual who chooses to lead is crucial to understanding how leadership styles and traits are established, and the potential impact these can have within various organizational settings. Ultimately, anyone choosing to lead is first and foremost a person with goals, ambitions, motivators, and ethics.

A program manager must always understand the environment, culture, and organizational issues that the program is to be developed in. Without this understanding, inherent risks, lack of communication, politics, and organizational fiefdoms can become a huge issue.

Recently, I was asked to lead a $30 million five-year program for a city government. The program had been started and dropped four times in the past, but this time it appeared to have some traction. Each of the agency directors had already signed a memorandum of understanding committing to the program conclusion, and a budget had been established to handle the long-term effort. However, as in every governmental environment, the implementation required a collection of skills including interpersonal relationships, formally requesting, arguing, and pushing to get the subordinate government officials onto the same page. Even with that, a single job description for a data technical person could take months of revisions, additions, legal review, and tailoring to the point that no one actually could ever meet the criteria being requested. The program progressed nicely with the project teams focused on the data, cleanup, quantification of issues, assessments of potential long-term resolutions, and technological repairs; but on the program management side, it was a constant battle to work with disparate groups to come to a common understanding even when doing so would make their individual offices run smoother and with less potential for error. The program manager is responsible for working within the organizational boundaries, overcoming barriers and wherever possible keeping the political infighting, posturing, and positioning out of the mix. Had we given in to the short-term conflicts that were taking place, the project teams would be shifted from one direction to another and never accomplish the goals set out for them.

Instead, a plan of action for the program team was created that enabled them to work mostly undisturbed while the program manager battled the ongoing issues and concerns. Only when a final decision was made that had a significant impact on the project would the program manager interrupt the work with the teams. But even at that stage, we still fell back on the tools created for the program and ran the change through the Program Control Board (PCB) to evaluate the change as to the benefits and costs to the program. In most cases, the board acknowledged the value of the changes but either deferred the change for a later implementation or defined it as out of scope for the initial efforts. Through this formalized and informal process of working with disparate stakeholders in a very distributed environment, the program made tremendous progress without significant interruption of new ideas, fiefdom buildings, or conflicting departments.

In the end, understanding the organization is critical to the success of the program. If "organizations breed anxiety in the individuals through the demands on individual performance, forcing individuals to work together performing tasks that are not enjoyed, and often treating people in impersonal and cold manners while requiring the same individuals to demonstrate loyalty and support for the organization itself" (Gabriel and Carr 2002, 355), [then…]. Anxiety is seen as an "incapacitating emotion which individuals defend themselves against through the mechanisms of defense" (Gabriel and Carr 2002, 355). The negative impact of the defense against anxiety in individuals is that organizations "often resort to dysfunctional routines which stunt creativity, block the expression of emotion or conflict, and above all, undermine the organization's rational and effective functioning" (Gabriel and Carr 2002, 356). These organizational defenses against anxiety can create collective delusions, causing an internal panic and flight from nonexistent threats while ignoring real issues.

Gabriel and Carr (2002) further posit that "excessive anxiety leads to highly dysfunctional defensive routines, while inadequate anxiety breeds complacency, inertia and gradual decay" (356). The program manager must understand these impacts and work to counter them through the creation of a positive and rewarding environment. In the example listed earlier, had the team members been required to participate in the conversations with all of the stakeholders, they could very well have become disenchanted with the effort and their ability to solve the problem. But with the program manager facing the stakeholders, the project team members were able to work in a much more comfortable working environment, thereby minimizing the negativity of the organization as a whole.

To be successful in a culture building high anxiety and stress for employees, a program manager must create a positive environment where team members are focused on delivering benefits to the organization through approaches that leverage their skill sets and keep them challenged. To foster this environment a program manager must be able to create clear communication channels as well as ensure that conflict resolution is a positive experience and that team members can work out differences without hostility or management intervention.

In the next parts of the book, we will discuss some very real leadership styles. While each has positives and negatives, there can often be reasons to use one or another to overcome a particular challenge or to assist a team member in moving forward. While you will immediately see some of the leadership styles as negative and ones that you would not want to use, please look at both the advantages and disadvantages to each. When I look at my own leadership style, I want to believe that because of my personality, the success that I have had in building HPTs, and my confidence in working with people in a positive and constructive way, I must be a positive leader and demonstrate all of the positive characteristics. To some extent this is more ego than fact. In looking over past engagements, I can quickly see situations in which I consciously chose to use a more negative approach either on an individual basis or when challenged by organizational and cultural issues. An effective program manager will purposely choose to use a combination of styles to accomplish the goals of the program and to drive team members to the highest level of creativity, innovation, and personal drive.

9

Leadership Theories

"A leader has to appear consistent. That doesn't mean he has to be consistent."

—James Callahan

To understand the effect the leadership style of a program manager can have on a program team, it is first necessary to understand various leadership styles. Each of the styles discussed here, although not a comprehensive list, has positive and negative qualities, and no single style will be effective in all scenarios.

Many leadership theories have evolved over the last 100 years ranging from great man, leadership traits, and behavioral to transformational, transactional, dark versus light, situational, and charismatic. Researchers have made efforts to link some of the theories across these leadership islands. Each model has its own pros, cons, assumptions, and limitations, but current research is much more focused on situational and transformational leadership styles, while the ongoing debate between born versus made leaders continues. Leadership gurus continue to offer new models as variations to many already existing models. Max Weber, MacGregor Burns, Bernard Bass, and Warren Bennis are a few of the important researchers in the area of leadership. Understanding the variances between leadership styles and traits is vital to leveraging the leadership theory in the workplace.

Leadership theories have an extensive history and abound with research and supporting materials. Leadership contributes to the development and innovation of new products, increased services through better communication, empowering staff, and discourse on risky issues (Howell and Higgins 1990). Although the general value of leadership is well accepted, the preferred style of leadership for specific scenarios is debated. Whether transformational versus transactional, charismatic versus pseudo-transformational, situational versus servant, or authoritative versus collaborative, leadership style has a tremendous impact on a team and its ability to perform effectively, positively, and innovatively in the attempt to meet or exceed business goals. With so many different leadership styles, which one is the best for program management and team empowerment in simple or highly complex environments? The answer, of course, is it depends . . .

Each opportunity faced by a program manager has a different set of team members, management culture, program challenges, environmental challenges, client issues, and cost-related problems. It would be wonderful if there were a single management style or trait that could be pointed to as "the management style that makes an excellent program manager." However, there are so many factors that contribute to effective leadership that styles will and should vary based on the team and the organization. The ability of a leader to vary his or her approach to the situation is referred to as situational leadership.

Leadership is motivating people to set aside their own objectives and for a period of time contribute to the successful outcome of a team effort. Situational leadership requires knowing what styles are available, when to leverage them, and how to tailor them to the situation at hand. With the assortment of styles and awareness of the positive and negative consequences of each style, leaders can focus on the approaches most suited to building HPTs, increasing communication, developing positive conflict management approaches, avoiding demoralizing staff, and establishing a safe and productive program management team.

Thomas Carlyle initially proposed the great-man theory, which was based on the belief that leaders are born and not made. In his opinion, a true leader was someone born to lead and who was naturally charismatic, intrinsically motivated, visionary, and empathetic. Although numerous examples could be cited to support this theory, including Winston Churchill, George Patton, Napoleon Bonaparte, and Alexander the Great, these same leaders were certainly born to greatness but many were highly educated in leadership throughout their lifetime. If we believe that leaders are born and not made through education, interaction, experience, and the challenges they face, then why invest in attending business school and seminars or reading leadership journals and books? If this theory was accurate, then leaders could just be placed in executive roles and not have to bother with all of the hard work of education, experience, and building knowledge. According to the great-man theory, education and experience would simply make bad leaders better and great leaders even greater.

The question of "Great Man," or born versus made leaders, has a deep history. In his 1915 book, *The Principles of Scientific Management*, Frederick Taylor also proposed that "captains of industry are born not made. In the future it must be understood that leaders must be trained right as well as born right" (2). Taylor referred to the implementation of scientific methodology as a tool for leaders, while still inferring that the leaders themselves must be inherently capable, with the proper mind, background, culture, education, and drive. Taylor also offered that "in the past the man has been first; in the future the systems must be first. This in no sense, however, implies that great men are not needed" (2).

LEADERSHIP STYLES AND TRAITS

Transactional Leadership

Transactional leadership, also known as managerial leadership, focuses on the role of supervision, organization, and group performance. This theory of leadership was first described by sociologist Max Weber (1946) and expanded on by Bernard Bass (1981). In Weber's interpretation, the average worker was motivated to achieve the best for him or herself and the organization. A leader needed to harness that intrinsic motivation and direct it toward the common goal of the organization. In other words, people will work harder if there is some form of personal gain or reward for themselves (see Figure 9.1). Whether that is additional salary, a bonus if the objectives are achieved, time off when the program is completed, or a potential promotion with the inherent benefits associated with it, personal rewards can generate very positive results. In later transactional leadership theory, the reward of the production/performance model was in line with Taylor (1915). In this model, employees will improve performance when rewarded appropriately, and therefore the production model is one of a single transaction—work to achieve goals and receive a reward for performance. Very little thought is given to the pay and benefits that the workers receive. Instead they see this additional effort outside their current workload and expect to be rewarded for it.

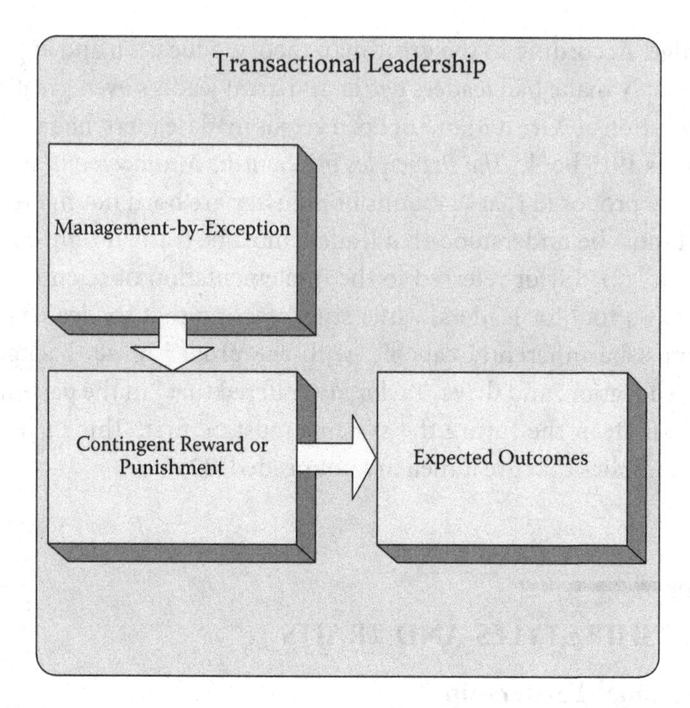

FIGURE 9.1
Transactional leadership techniques.

Transactional leadership is generally short term in nature and based on the immediate deliverables for the task at hand. An employee who delivers the requisite production within the established time frame is rewarded, while those who do not achieve are penalized. This form of leadership is based on a system of rewards and punishments and follows a series of assumptions regarding employees and team members:

- Team members perform their best when the chain of command is definite and clear.
- Employees are motivated by rewards and punishments.
- Individuals need to be carefully monitored to ensure that expectations are met.
- Followers obey the instructions and commands of the leader.

Transactional leadership is often used in simple business models such as retail, food services, production facilities, and manufacturing: when employees are successful, they are rewarded; when they fail, they are reprimanded or punished. Usually, this leadership style has been most successful in functional, operational, low-paying positions. Workers such as hourly employees in operational teams (road crews, builders, assembly lines, etc.) are good candidates for transactional leadership approaches.

In transactional leadership, rewards and punishments are contingent upon the performance of the individual. The leader views the relationship between managers and subordinates as an exchange—you give me something for something in return. When subordinates perform well, they receive some type of reward. When they perform poorly, they will be reprimanded or punished in some way, such as job loss or being passed over for promotion. Rules, procedures, and standards are essential in transactional leadership. Individuals in a transactional relationship are not encouraged to be creative or to find new solutions to problems. Long-term personal growth is not the intent; instead the focus is on delivering according to the defined rules and objectives. Research has found that transactional leadership tends to be most effective in situations where problems are simple and clearly defined and is least effective in situations where creativity, innovation and long-term benefits are encouraged. While transactional leadership can be effective in some situations, it is generally considered an insufficient and ineffective leadership style that results in morale and personnel issues later on. Transactional approaches often prevent personal growth and the achievement of individual and team potential, creating a negative environment for individuals who are intrinsically motivated and those looking for personal growth.

Transactional leadership focuses on the role of supervision, organization, and group performance. Based on the research of Taylor (1915), "the average workman must be able to measure what he has accomplished and clearly see his reward at the end of each day if he is to do his best" (94). There is a level of motivation for workers that can benefit from the application of transactional leadership.

Transactional leaders are those who:

- Approach followers with an eye to exchanging one thing for another
- Pursue a cost-benefit, economic exchange to meet subordinates' current material and psychic needs in return for "contracted" services rendered by the subordinate
- Follow a very self-motivated and self-rewarding style of leadership
- Will often take credit for team successes
- Often blame failures on the team, tools, or clients rather than accepting that as a leader he or she has failed to meet objectives

- Often create a negative environment with team members competing against each other for the leader's approval
- Leave conflict to the team to resolve, often resulting in negative conflict between team members
- Make arbitrary decisions and mandate implementation
- Demonstrate a different behavior and communication style with management than they do with the team
- Are often self-centered and egocentric

Now, a transactional leader who demonstrates these traits can have very negative results over the long term, but the approach can also be leveraged positively in an environment where teams need to be driven until they can drive themselves or where the process is repeatable in nature and leverages more manual-based labor. A program manager may choose this style when there is a great resistance to change, refusal to accept new process, or resentment of a new manager. The Hawthorne study (Elton 1949) demonstrated that employees can often be motivated simply by recognizing that their performance is being monitored and watched. Even the slightest form of metrics can cause behavior changes to team members, pushing them toward being a more successful group. In this study a group of workers were informed that they would be measured on performance but were not told what measures would be used. Before researchers could begin the study, the workers increased productivity, dramatically indicating that just the awareness of measures being put in place has an effect on how workers perform.

We will discuss this further in the section situational leadership, but the approach taken by a leader is often a conscious decision made to drive the team toward greater success. It is when leaders demonstrate these traits unconsciously and do not understand the impact that negative results can occur.

A relationship between two people is based on the level of exchange they have. Exchange need not be money or material; it can be anything. The more exchanges between two individuals, the stronger the relation. Your manager expects more productivity from you for greater rewards. In this way, if something is done based on the return, then that relation is called "transactional." In politics, leaders announce benefits in their agenda in exchange for the vote from citizens. In business, leaders announce rewards in return for productivity. The previous examples of relationships are all about requirements, conditions, and rewards (or punishment) and referred to as transactional.

In program management creativity, innovation, and teamwork achieve success. The relationship between a program manager and the team is more of an interactive, guidance-based one than one of a single transaction. The leader needs to drive individuals to perform as a team that together produces more than the individuals can achieve on their own. Therefore the program manager applies leadership techniques that encourage and empower team members to work collaboratively and that limit the amount of management interaction wherever possible. The team functions as a single unit, understanding individual roles and responsibilities and interacting to achieve objectives. Transactional leadership would drag a team down to compete for rewards on an individual basis or to avoid penalties, creating an adverse culture of unique individuals. Leveraging transactional leadership in a team long term will work against the team mentality and generate internal strife. However, leveraging this approach on an individual or situational basis to overcome an obstacle or to push a team member can be advantageous. The choice to use this style should be a conscious one that is tailored to the situation/individual and is changed once the objective has been achieved. Program managers need to focus on building teams and driving performance, and a reward/punishment approach generates a contradictory individual-first environment.

Transformational Leadership

> You don't lead by pointing and telling people some place to go. You lead by going to that place and making a case.
>
> —**Ken Kesey**

A true leader is one that is capable of rising to the challenge of the environment, the team, and the established objectives. The leader must be able to motivate in multiple scenarios, respond to challenges of various organizational entities, and recognize the tools available and leverage them in an effective manner.

The most current views of management style show two types of leadership on different ends of a spectrum. As shown in Figure 9.2, transactional leadership is on the left and transformational leadership is on the right. No one leader is considered to be 100 percent transactional or

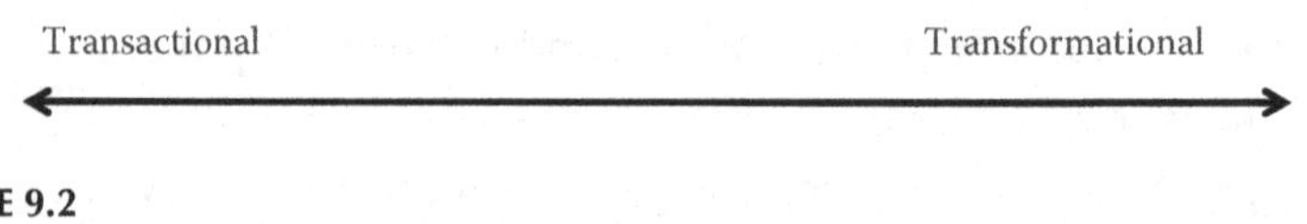

FIGURE 9.2
The leader spectrum.

transformational; instead all leaders are some shades of gray between the two.

On the opposite end of the management spectrum from transactional leadership is the more charismatic, inspirational, and intellectual leadership most often referred to as transformational. A transformational leader motivates and empowers employees and teams by creating common visions driving them toward intrinsically motivated success. Those teams are driven by the achievement of objectives, can see the big picture for the organization, and understand their role in it, working most comfortably when provided with all of the program details. Teams such as these are wonderful candidates for transformational leadership approaches. Most people agree that the advantage of the charismatic/transformational leader is that he or she provides an environment that empowers the staff, drives creativity, increases communication, and generates more innovation as well as reducing attrition rates and eliminating team conflicts.

Again, when evaluating this on a continuum, with transactional on the far left and transformational on the far right, it is important to recognize that there are many shades of gray in between. The environment of the operation, type of worker, educational level of average staff members, and socioeconomic classification of all levels of staff will all contribute to adjustments on the scale.

While transactional leadership is understood as a base form of management, there are many occasions where we see a completely different kind of style implemented in personal life. Quite often we will see a mom devoted to her child. Mom doesn't expect anything from the child, and the service she provides in raising the child is unconditional, dedicated, and committed. Mom plays a major role in shaping the child's future life. This type of relation is referred to as transformational in that the mother is focused on the betterment of the child with no ulterior motive or expected reward, other than to raise a happy and healthy child. Transformational leaders work toward a common goal with followers; put followers in front and develop them; take followers to the next level; and inspire followers to transcend their own self-interests in achieving superior results.

The most common ways to recognize transformational leaders are situational. A transformational leader:

- "Recognizes and exploits an existing need or demand of a potential follower . . (and) looks for potential motives in followers, seeks to satisfy higher needs, and engages the full person of the follower" (Burns 1985, 4).
- Recognizes the transactional needs in potential followers "but tends to go further, seeking to arouse and satisfy higher needs, to engage the full person of the follower … to a higher level of need according to Maslow's hierarchy of needs" (Bass 1985, 20).
- "Facilitates a redefinition of a people's mission and vision, a renewal of their commitment and the restructuring of their systems for goal accomplishment. It is a relationship of mutual stimulation and elevation that converts followers into leaders and may convert leaders into moral agents. Hence, transformational leadership must be grounded in moral foundations"(Leithwood, as cited in Cashin et al. 2000, 1).

And the transformation leader must both understand and accept the cultural environment of the organization: "Transactional leaders work within the organizational culture as it exists; the transformational leader changes the organizational culture" (Bass 1985, 31). In other words, the culture of the organization is understood, but the necessary changes to drive a transformational environment are not lost on the leader. Instead of just focusing on the team he or she is responsible for, the transformational leader manages both up and down the chain of command driving change in both directions. Table 9.1 shows the differences between transactional and transformation leadership.

Early studies portrayed transactional leadership and transformational leadership as mutually exclusive, but many researchers today view them as on a continuum rather than as opposites. The transformational leadership style is complementary to the transactional style and likely to be ineffective in the total absence of a transactional relationship between leaders and subordinates. There are personnel-related situations where any leader will need to use a transactional approach to resolve problems in performance or work ethics. As such, both styles are necessary for an effective program manager to leverage, but the style chosen must be conscious and proactive based on the situation at hand.

TABLE 9.1

Transactional versus Transformational

Transactional Leadership	Transformational Leadership
Leaders reward or punish based on the accomplishments of the team member.	Leaders motivate followers through an emotional response to contribute to a set of objectives that are outside of a transactional interaction.
Leadership is reactive to situations, and it deals with present issues rather than long-term objectives.	Leadership is proactive and forms new expectations in followers, driving intrinsic motivation and innovation.
Team members respond to instructions rather than achieving objectives through personal drive and ambition.	Transformational leaders are charismatic and empower, inspire, and stimulate followers to achieve group-based goals and objectives.
Leaders motivate by defining short-term goals and promising rewards for desired performance or by threatening for nonperformance.	Leaders empower followers to innovate, problem solve, and increase personal growth to achieve team objectives.
Leadership depends on a power base that can reward or penalize subordinates for their performance.	Leaders demonstrate and communicate visions that followers can fulfill and leverage interpersonal skills to develop emotional bonds with followers.
Leaders use carrot and stick to drive employees to success, rewarding positive behavior and punishing negative behavior.	Leaders motivate and empower followers to work toward goals that go beyond personal self-interest.

The following is an example of these traits. Very early in my career a programmer told me that he had deleted 80,000 records from a production database. I immediately contacted the client and told them the situation, asking that they stop working on the system while we solved the problem. Then I went to our database administrator to see what could be done. By leveraging the transaction log, we were able to restore the deleted records, and the client was back up and running within an hour.

Unfortunately, the CEO of the firm differed in his management style from the one I had adopted and called me, insisting on the name of the programmer who made the error. I explained that it happened during my watch and therefore was my responsibility. Although he pressed for the name of the person, I refused to give any name other than my own.

While I had unintentionally created a transformational leadership environment where mistakes were allowed, communication between team members increased, and risks were openly discussed. The individuals on the team were empowered to make decisions and to leverage creative processes in overcoming obstacles. As such, the team grew together, effectively resolving conflicts in a positive and proactive manner and developing a continuous self-improvement style. As a whole, schedules were met, program completion reached its highest levels, and we were able to proactively respond to risks and issues, exceeding stakeholder expectations.

Weber (1947) proposed a set of theories of transitional leadership in his model of transactional and transitional leadership authority. The foundation of the more advanced theories of transitional leadership is Weber's (1947) model of charismatic/hero (transformational), bureaucratic (transactional), and traditional (authoritative). Weber described these three frames and founded a generation of research and analysis on the topic. From his initial interpretation, the transactional/transformational model has been discussed in numerous research papers, books, methodologies, and philosophies. Although charismatic leadership has foundations in the great-man theory, the transformer role is one that has continued to evolve beyond the individual to models and frameworks that can be taught.

Bass (2004) created a tool called the Multifactor Leadership Questionnaire (MLQ), which is used to measure the level of transformational versus transactional leadership style. In this tool, most leaders score both transactional and some transformational style traits depending on the environment and situation they are currently working in. But it is clear from the traits of a transactional leader that the team will quickly move toward a more externally driven motivation rather than an intrinsic desire to achieve objectives. Therefore the use of transactional leadership styles is a somewhat dangerous approach and is generally leveraged for only a short time to drive the team toward using new standards, accepting new leadership, and achieving a common vision.

Based on detailed studies of various research done on transformational leadership, the broader traits of a transformational leader include but are not limited to:

- Has a clear sense of purpose, expressed simply
- Is value driven (e.g., has core values and congruent behavior)
- Is a strong role model
- Has high expectations
- Is persistent
- Is self-knowing
- Has a perpetual desire for learning
- Loves work

- Is a lifelong learner
- Identifies him or herself as a change agent
- Is enthusiastic
- Is able to attract and inspire others
- Is strategic
- Is an effective communicator
- Is emotionally mature
- Is courageous
- Takes risks
- Shares risks
- Is visionary
- Is unwilling to believe in failure
- Has a sense of public need
- Is considerate of the personal needs of an employee
- Listens to all viewpoints to develop a spirit of cooperation
- Is mentoring
- Is able to deal with complexity, uncertainty, and ambiguity

This list is also the same set of characteristics that make up a successful program manager's portfolio. We look to program managers to drive programs and benefits through a level of self-actualization and confidence that can also be translated into tangible and achievable goals and objectives. A program manager needs to create a program, delivering a set of benefits within a limited time frame and budget, while ensuring that the effort maintains ongoing alignment with strategic objectives; manages risks, benefit realization, and stakeholder expectations/involvement; and ensures overall governance over activities. Obviously, effective program managers cannot achieve these responsibilities alone; they need a team of empowered, motivated, self-managed, and driven followers who understand and agree with the vision and benefits for the program. The larger the program, the larger the team and the more unique personalities involved. Program managers are often faced with situations that other managers would deem impossible, and therefore they need a team capable of supporting the objectives and able to make day-to-day decisions that will be supported by their leadership. The traits listed previously and in Figure 9.3 demonstrate the personality and management style of effective program managers.

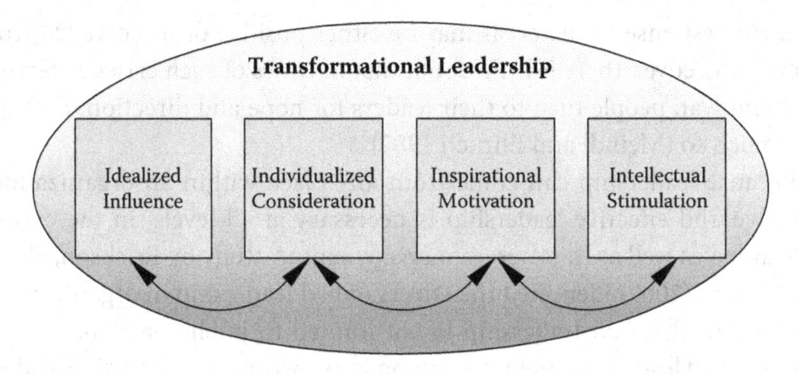

FIGURE 9.3
Transformational leadership traits.

Light and Dark Leadership

> The first responsibility of a leader is to define reality. The last is to say thank you. In between, the leader is a servant.
>
> **—Max DePree**

Gallagher (2002) proposed that bad leaders are a critical factor in organizations and have a negative impact on the success of initiatives: "We have effective leaders, we have strong leaders and good leaders but we also have ineffective leaders, weak leaders, and of course bad leaders" (27). Leaders accomplish goals that could not otherwise be done by individuals through motivation, communication, conflict resolution, and vision. Positive leaders motivate through positive motivational techniques and leverage drivers such as reward, incentives, and positive reinforcement, while other, more negative leadership is done through fear, intimidation, ridicule, and hostility. Tyrants can often be considered as leaders who drive advances through fear and intimidation; "the traditional emphasis on effective leadership, strong leadership, good leadership, visionary, and inspirational leadership, raises the philosophical dilemma of what constitutes ineffective leadership, weak leadership, bad leadership, non-inspirational, and non-situational leadership" (Gallagher 2000, 27).

To find a solution to the "meltdown of corporate ethics" we have a distinct need for positive leadership and authentic leadership development, as evidenced by the highly publicized scandals involving Enron, WorldCom, Arthur Andersen, and Adelphia (Luthans and Avolio 2003, 241). In today's economy, crises are commonplace and should be expected because every organizational leader will face at least one crisis during his or her tenure and the response to the crisis may be either positive or negative (Mitroff 2005). Moreover, there is little doubt that in times of such crises as terrorism and war, people turn to their leaders for hope and direction, perhaps too much so (Meindl and Ehrlich 1987).

Because leadership can come from any place within an organization, positive and effective leadership is necessary at all levels, in the executive arena as well as in program management. A study by Erickson, Shaw, and Agabe (2007) identified the effects of bad leadership in organizations and found that bad leadership is not limited to public or global crises; rather, "bad leaders are not uncommon in the workplace" (39). Bad leaders can often be found at any level of an organization and are recognized as those who "consistently failed to motivate and appropriately reward staff; . . . they did not professionally develop staff, recognize them as people, or understand their strengths and weaknesses" (38). To be effective, leadership must be a combination of mutual and reciprocal exchanges between leaders and followers. Both leaders and followers working together need to understand the value of the outcomes, participate in two-way communication, and agree on a common set of goals (scope) for the initiatives before them.

In many cases, bad leadership does not have a major impact on the organization as a whole, yet "when a leader is perceived as bad by one member of an organization, that opinion is perceived to be widely held by other members of the organization" (Erickson, Shaw, and Agabe 2007, 37). Therefore bad leadership can have a cumulative impact. Because effective leadership is so vital to an organization, if leaders are not selected wisely, organizational damage, attrition, morale issues, and productivity losses can quickly arise.

While extreme environmental factors undoubtedly influence potential organizational results, numerous studies have shown that leaders have a positive impact on important outcomes (Reichard and Avolio 2005). Furthermore, because of the escalating complexity of the organizational environment, leaders are needed at every organizational level. Now we find that continuous leader development is becoming a strategic priority for many organizations (Day, Zaccaro, and Halpin 2004). As a program manager, you may be called upon to support the organization in the event of a crisis and to assist in managing your staff to still achieve strategic objectives for the organization, decrease fears, and eliminate rumor mills. It is quite common for a program manager to walk into an environment where bad leadership has had a tremendously negative effect on the staff and the initiatives of the organization. As such, the program manager will need to enter the opportunity with a clear set of ethical and valued skills, empowering staff and, at least within the program, building a more healthy and effective culture.

Bass (1999), Howell (1992), Ciulla (2003), and Conger and Kanungo (1988) all refer to leaders who operate outside of ethical boundaries as unethical leaders, unethical charismatics, and pseudo-transformational leaders. In new studies these labels have been consolidated and used interchangeably in the definition of dark leaders.

Dark leadership refers to the inner urges, compulsions, motivations, and dysfunctions that drive people toward success regardless of ethical consequences (McIntosh and Rima 1997). The study of the dark side of personality among leaders plays an important role in helping organizations identify those with the potential for derailment, deviant behaviors, and poor work performance (Khoo and Burch 2008). While dark leaders have a long-term negative effect on an organization, and those who are unconsciously aware of the fact that their leadership style is considered to be unethical are disastrous to an organization's culture, there are times when a program manager may have to leverage this approach to move a team forward against the grain of the organization and what has been considered to be acceptable behavior. Program managers may choose to leverage the approach, focused on transitioning over time to a more positive end result of ethical and effective leadership. Situations where the culture has reached a point at which ethics and morals are questionable may force a program manager to impose his or her own set of ethics and morals, driving the team toward what will eventually be a positive and ethical environment.

Dark leaders are perceived as those who push from an unethical perspective regardless of the consequences of their actions. However, viewed from another perspective, if the organization itself is unethical, a dark leader would be one pushing teams to work toward success regardless of the culture and morals of the firm. In this case, the program manager would be seen as "going against the grain," not being a team player, and being a threat to the organization; but from the leader's perspective, he or she would be pushing the team to greater success regardless of the consequences. This struggle is more common than you may think. Organizations will often reach a point of becoming obsolete based on management practices, poor service, and poor quality standards. To resolve this scenario a program manager may be brought in who is capable of driving toward a more effective service delivery model. Yet the negative culture is so ingrained and accepted, the program manager may choose to demonstrate some of the traits of a dark leader to drive toward a more ethical and positive environment.

The importance of leadership in shaping an organization is best demonstrated through evaluation of ethical and social norms within the organizational environment. Leadership drives success, which in turn affects the organizational standards. Most firms will not resist what is successful regardless of the history and culture they have established. However, be warned that if you push back against the organization, you can generate detractors and resisters, all of whom will attempt to undermine your efforts to succeed.

Whereas light leadership is morally driven and contributes to both the leader's and the organization's effectiveness (Engelbrecht, Van Aswegen, and Theron 2004), dark leaders expend a tremendous amount of organizational resources in the pursuit of personal visions instead of organizational strategic objectives, which threatens the long-term success and viability of the organization. As such, assuming an organization is an ethical one following a positive and moral path, having a dark leader could result in compliance, ethical, legal or financial issues, and is likely to cause issues with communication, loyalty, motivation, and attribution, all of which are necessary for organizational success (Conger 1990).

McIntosh and Rima (1997) have identified five types of dark leadership:

1. Compulsive
2. Narcissistic
3. Paranoid
4. Codependent
5. Passive-aggressive.

Each of these has detrimental personality aspects that cause them to drive toward success often following unethical behavior and are perceived as unethical.

Compulsive leaders focus on the need to maintain absolute power over all aspects of the organization. These individuals see their leadership as an extension of themselves and their personal lives and therefore attempt to exert control over every aspect of a project or effort. Compulsive leaders fear anything outside their control and attempt to control anything that can impact their efforts.

Narcissistic leaders often come from the classic narcissistic personality type. As such the narcissist is motivated to achieve the respect and admiration of others and especially their superiors (Judge, LePine, and Rich 2006). While narcissism is often related to charismatic leadership and can be one of the aspects that a leader leverages in his or her willingness to lead, Khoo and Burch (2008) suggest that the relationship between charisma and narcissism is not a predictor for effectiveness. Instead, narcissists focus on their personal goals and ambitions, which are generally accomplished at the expense of others through approaches such as aggressive, derogatory behavior, and aggrandizing personal accomplishments. A major difference is that the narcissistic personality lacks empathy, one of our key traits of leadership (Morf and Rhodewalt 2001).

Paranoid leaders live in a constant state of denial. Although often quite brilliant, paranoid leaders are those who are convinced that others are actively working against them to take away their efforts, leadership, or achievements. These leaders are generally insecure in their own abilities and are jealous of the abilities others generate, sometimes to the pathological point. Similar to the compulsive leader, a paranoid leader will attempt to control all aspects of a program and can become hostile toward anyone trying to offer assistance. The only effort a paranoid leader will support is one that he or she can fully control. Success for this leadership style is often based on the leader's talents and can be undermined by team members out of frustration and bad morale (Williams 2005).

Codependent leaders are those who attempt to take control of and responsibility for others, and blame their ineffective actions on others (Goff and Goff 1991). The inclusion of a codependent leader in a healthy organization can be destructive. They can also get caught up in unhealthy behaviors and addictive diseases such as alcoholism, drug addiction, and social avoidance (Larsen and Goodsen 1993). The workplace behavior of a codependent leader is often destructive to an organization and can be demonstrated through non-work-related efforts, poor organizational skills, ineffective conflict resolution, inefficiency, lack of interpersonal skills, inability to motivate workers, and failure to communicate a consistent, concise vision (Schaef and Fassel 1998).

Passive-aggressive leaders demonstrate a set of personality traits, often conflicting, where on one hand they actively avoid social contact and complain of others not understanding them along with personal complaints of unfair personal misfortune, while on the other hand they are often standoffish, hostile, defiant, argumentative, sullen, passively deceitful, obstructive, scapegoating, resisting change, protecting turf, and unreasonably critical of authority (Johnson and Klee 2007). These behaviors are known to cause low levels of respect and organizational issues with team development and effective leadership.

These five personality traits (narcissistic, codependent, paranoid, passive-aggressive, and compulsive) can work against organizational culture as well as the necessary leadership and training to facilitate team development long term, but can be effective in short-term circumstances to motivate a team or to push them out of their comfort zone. While the negative aspects of a paranoid or narcissistic leader are somewhat obvious, they can also be used as tools to drive teams forward through an authoritative command-and-control process.

Program managers may choose to adopt one style or more based on the immediate needs of the team and the program. Of course maintaining these styles over the long term can be detrimental to the morale and effective communication of the program members and should therefore be used carefully and consciously. Short-term and intentional use can have a positive effect moving program teams to reject the organizational culture and move toward one that is more positive for the program needs.

In one program I ran, the team had become so inured to failure that even discussing success was taboo. There was a litany of excuses and reasons why success was not possible: "Design won't give us anything on time and will change the requirements repeatedly"; "The features will change midstream"; "Maintenance issues will take precedence"; "We don't have a voice in the decision making"; "We don't have the skill set or technology knowledge to meet the needs"; "No one will buy us tools to do our job"; and on and on. While many of these concerns were valid, I chose to use a narcissistic approach to solving the program issues. "No one tells me no" was my motto to all, including design, executives, finance, etc. I grouped the team to establish a project plan that would achieve success, and we held the other agencies' feet to the fire to achieve success.

If design couldn't make its goals, the problem was taken to the executive board and pointed out that the delays coming from design were preventing program success. When scope creep became an issue, we went to the sponsors and said you can have what you asked for in the time frame and cost that you established, or you can have this yet undefined scope increase, but based on schedule, cost, and quality, both cannot be done. It was a very narcissistic approach from a subject matter expert (SME) and program manager perspective and did not make many friends in other departments, but the team started to show energy and innovation, even some creativity, and the program was successful in realizing the defined benefits.

There was one meeting with the project managers where the quietest individual on the team finally spoke up. "You know," he said, "if we combine these two efforts with a bit of modifications and leverage the combined budget, we could give the sponsor what they want without schedule or cost change." Finally, we had a team that was looking outside the box to identify strategies to be successful instead of giving up the battle before the first shot was fired. And it was time to change my management style to a more proactive and supporting one.

Command and Control

> Your position never gives you the right to command. It only imposes on you the duty of so living your life that others may receive your orders without being humiliated.
>
> **—Dag Hammarskjöld**

Organizations can have either a strong or weak culture. Participants in a strong culture generally demonstrate a clear alignment with the organizational values and willingly contribute to the common vision, while participants in a weak culture must be more formally controlled usually through transactional or command-and-control management styles, and driven to achieve objectives through process, bureaucracy, and authority. Whether weak or strong, an organizational culture can be evaluated through various methods and tools that have been used extensively in research.

Hofstede (1980) identified four dimensions of culture including:

- Power distance – the degree of separation demonstrated between levels of power or authority within the organization
- Uncertainty avoidance – the extent to which the culture accepts or resists risk (risk tolerance level)
- Individualism versus collectivism – the extent to which people are able and expected to stand up or to operate as part of a team

- Masculinity versus femininity – the value placed within the organization on traditional gender roles such as competitiveness, assertiveness, ambition, and the accumulation of wealth

These four dimensions provide a framework for analyzing and classifying the organizational culture.

Another perspective on this is the four types of culture (Handy 1985):

- Power culture – places the power among only a few within the organization with a central group or person driving the agenda and purpose or the entity as a whole
- Role culture – formally delegated authority structure generally operating within a greater hierarchy with power defined by the title of an individual
- Task culture – teams formed and used to solve particular problems and generally lasting until a resolution is found or a task is completed
- Person culture – where each individual believes that he or she is equal or superior to the organization itself, and each member contributes what he or she feels is needed by the organization such as a professional partnership

These four levels are often referred to as command and control and function much like the military. Many of my friends who have a military background swear by this approach in handling all sorts of management challenges. These individuals are very comfortable with the chain of command and following instructions from the leader downward. With a great deal of effort, I have tried to shake their belief structure but to no avail. In the programs that they run, they continue to follow the chain of command and find it to be quite effective, possibly because many of their team members also were in the military.

Personally, I find the environment a bit stifling as each team knows only what upper management determines they need to know. Teams don't work as a cohesive unit and don't contribute to each other's success, and programs are difficult to manage as each team sees themselves as a unique unit rather than part of a collective organization delivering benefits to the organization. A network team may build a network to the specifications provided but may not be aware that a change in the programming due to a realized risk has now shifted the focus from a client/server to a service-oriented architecture. While the initial network meets the standards that were provided, the shift on the development side leaves it ineffective as a solution. As a program manager, my goal is to make as many people as possible aware of the risk, opportunities, and new approaches to ensure that we make the most of each change and fully integrate knowledge across the program. Furthermore, team members should be comfortable in coming to the program manager regardless of their concern or to which project they are assigned. Proper communication establishes an environment where each person with an issue, regardless of its foundation, should feel comfortable in communicating that issue and be listened to and considered. Of course I am not saying that all concerns are valid, just that the communication channels should be open and that when concerns are vocalized, we learn more about the teams, projects, management efforts, and risks of programs.

Situational Leadership

> Management is efficiency in climbing the ladder of success; leadership determines whether the ladder is leaning against the right wall.
>
> —**Stephen R. Covey**

Situational leadership is one of those controversial topics with critics believing that the leader should be able to lead any team with the same style of leadership using positive reinforcement, rewards, motivation, and building team morale, and supporters believing that in reality program managers face various environments all with a predefined culture that has been established and accepted over the long term. In some cases, to meet our goals we have to consciously move out of our comfort zone of leadership style and implement practices that will motivate and push the teams toward the goals of the program.

The concept of a "situational manager" was initially proposed by Blanchard (1982) when he evaluated a number of leaders in various environments. While the leaders would gravitate toward a particular style naturally, many were consciously choosing a different style outside of their comfort zone. These managers were more effective than their counterparts because they tailored their leadership approach to the organizational culture and environmental factors. The theory of situational leadership is based on the assumption "that one size does not fit all. Only by reviewing the situation you are in—incorporating the work environment, followers, and industry challenges—can you best determine the leadership behaviors that would make you the most effective" (Topping 2002, 3).

Leaders, but especially program managers, often find themselves in new and very different environments. Organizations that have a destructive, negative, toxic, or hostile environment can be places where individuals often feel that their position, reputation, or job is in jeopardy, causing a negative set of behaviors and a level of risk aversion that can threaten innovation, communication, and creative efforts. Team attrition, morale, and personal job satisfaction are all of great concern in these environments. Program managers in an environment such as this can find themselves at risk for challenging the status quo and attempting to drive success through a more positive leadership style. They will need to understand these factors and work to foster an environment that is positive regardless of the corporate culture, creating a positive team-based bubble within the organizational environment. As such, their normal management style may not be effective. Situational leadership involves identifying the organizational challenges and adjusting, often away from a preferred leadership style into one that is much more effective for the objectives of the program.

Conversely, organizations that reward success, provide positive feedback, and support innovation can often create more positive psychological impacts, such as workaholic staff, overachievers, and friendly competition between managers, departments, and leaders. In this kind of environment, program managers will often find that their comfort zone is more effective and they can start leading through a transformational style, expecting that users are self-motivated, intrinsically driven, and focused on the objectives of the program effort. Project managers may have already adapted PMI best practices and may be running projects with earned value analysis (EVA), risk management, and planning procedures. This obviously makes the program manager's responsibilities much easier as everyone starts with a common set of standards and a clear understanding of how projects and programs are managed.

Many researchers have started to focus more on organizational or situational management styles and impact. A leader who has been widely successful in one organization may find that he or she struggles dramatically within another (Drucker 2001). The leadership ability of an individual is often also correlated with his or her ability to read, respond, and adjust according to the environment that he or she encounters (see Figure 9.4). Evidence from previous studies of leadership leads us to believe that the

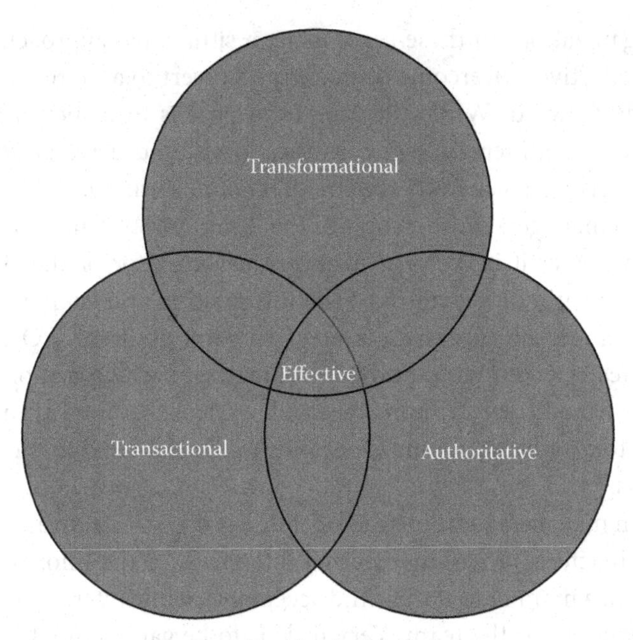

FIGURE 9.4
Situational leadership.

actual leadership employed not only must align with the immediate needs of the effort; it also must be appropriate to the situation.

The responsibility for understanding the environment and organization falls squarely on the shoulders of the program manager. It is his or her duty to do an analysis of the environment, study the impact that environment has had on its employees and past programs, and determine where the negative (and positive) environmental factors have come from. Doing a root-cause analysis on environmental and situational aspects of a firm will assist the program manager not only in understanding the environment but also in better determining the style that would be most effective in the situation.

Situational management has its detractors; for example, there are those who believe that leaders should always follow their beliefs to be successful and that remaining true to their beliefs will make programs successful regardless of the environment they are working in. However, my experience is that if program managers do not take the time to understand the environmental and situational factors, they will be ineffective in responding to very real concerns of their staff members and often very short lived in the organization. In these cases using a situational approach allows a leader to effectively overcome obstacles and revert to a more comfortable style going forward. While the long-term goal is to empower the staff, drive innovation, increase risk tolerance levels, and drive program success, the short term generally requires a bit of politicking.

Program managers must recognize that the environment they are operating in, whether it is the organization they work for or the client they are working with, drives much of the information and support they will receive. It is rarely a successful strategy to point out to a CEO that he or she operates as a command-and-control manager, which demoralizes the team. Nor would a program manager tell a client that their transactional approach to driving programs is negative to the team and to the potential for success.

Program managers must instead operate between the team and the environmental factors, providing external interfaces for the information that they need in a manner that they find acceptable, while filtering or restating the information for the team. Very little is to be gained in telling a team member that the client made derogatory comments about that person or the likelihood of the program's success. Program managers will hear a litany of complaints about individuals, the capabilities of the team, and the potential success of the effort. An effective program manager will listen to these concerns, internalize them, and watch for any of them to be realized with potential strategies to overcome them if they are accurate. For the success of the team, the program manager should work both with the client to dissuade them from expressing negatively to team members to ensure that the team members are receiving the support they need to be successful in achieving the program goals. This will ensure that the team is on the same page with the vision, challenges, and objectives for the program without being inundated with sponsor and one-off requests detracting from program goals. In addition, program team members can focus on the work at hand without being concerned about stakeholder impressions. From this, a program manager can build a positive and innovative environment. This impact is one of the key reasons for a stakeholder engagement plan and a stakeholder register to state who deals with whom and how such situations are best handled.

Sound a bit schizophrenic? The answer is yes. Initially, a program manager will deal with a number of political agendas, impressions, and historical experiences. But the program is a new one deserving of its own set of standards and experiences. A truly effective program manager will initially operate and be seen differently by every group with whom he or she interacts. The goal is that over time each group is provided up-to-date and honest information based on the program status and achievements. However, spending time overcoming impressions and negative history is ineffective and detracts from the work at hand. Let the program success speak for itself. Each group has a different agenda, political beliefs, concerns, risks, and availability to discuss issues; therefore the conversations and focus will differ. Over time, an effective program manager will be able to drive conversations and meetings to a more positive note, increasing communication and demonstrating success, but will not be taken seriously if he or she simply walks into efforts that have failed in the past and promises success.

This is actually an opportunity for program success and not a drain on program managers. They are often given information that outlines why other programs have failed, what client or management expectations were, how communications should have been handled, and where risk threshold levels are. An effective program manager will document all of these factors and will come up with plans to meet the expectations of the external stakeholder community and to ensure that mistakes of the past are not repeated.

However, the one factor that program managers can control is the interaction they have with the team. Program managers work not only with project managers as their direct reports but also with every staff member and contractor who report to the project manager. They must consistently demonstrate a level of confidence in the team, work to resolve conflicts in a positive way, and empower team members to drive innovation and achieve success while maintaining schedules, benefits, costs, and quality. They must consistently communicate the long-term vision, outline how each project contributes to benefits that will achieve the vision, and ensure that team members all recognize the value of the contribution they make. In addition, the communication plan for the program must be clearly outlined so that project managers, quite often focused on their specific project, can speak to the long-term program and have a method by which requirements outside of the project scope can be escalated to the program team for review and possible inclusion in the longer-term vision.

I have often worked with organizations that had failed programs, and my task was to take the remnants of the program, pick up the pieces, and attempt to develop strategies to avoid past mistakes and overcome obstacles from cost and schedule to staffing issues. One specific situation comes to mind where a program had been started three times in the past only to fail before it actually made any progress. This program had a wealth of information from previous contractors (all of whom seemed to feel that their responsibility was to report on the program in no less than 150 pages for each document they produced). By reading through the documents produced by past contractors and interviewing various stakeholder groups, a number of factors came to mind. The previous contractors treated the effort as a simple project that would deliver a unique product or service, potentially meeting the needs of one group of stakeholders but not others. They also did not work to achieve buy-in from multiple stakeholders in advance. Instead they took a large stakeholder community and tried to force that agenda on others—in this case other departments within the same government all with defined areas of responsibility. Finally, they did not establish open communication plans where all stakeholder communities were recognized for their contributions.

To restart this effort, we first defined it as a program and not a project, with multiple component efforts that would be managed together to achieve benefits not available from a single effort. We developed documents such as a charter, communication plan, risk management plan, and change management plan. A program steering committee was created and kick off meetings held with the entire stakeholder community. Then an initial project manager was assigned to drive the first project forward. This project manager was tasked with introducing the project, developing project management plans in alignment with the program management plans, and describing how the project would be a small piece in an overarching program, the benefits of which affected all stakeholders and provided long-term benefits to each group. While the initial project was bound by a specific scope statement, additional needs were collected by the program manager to collate them into a comprehensive list of tasks from stakeholders. Not every requirement would be done by the program, but many drove the creation of new projects that would increase the value to the stakeholder community.

Finally, communication and commitment were required for the stakeholder community. As the assigned program manager, my responsibility was to identify and evangelize the long-term benefits of the program and to assist stakeholder communities to better understand how their needs would be met in benefits realized throughout the program. The long-term road maps of benefits were identified, but as living and breathing documents. As more information became available, the benefits were adjusted and communicated outward, ensuring that stakeholders all maintained the same set of expectations for both benefits and the timing of realization as well as the costs and risks associated with them.

The key to success on a long-term effort like this is the communication of benefits as they are realized. Since the program would take up to three years to complete, reminding stakeholders of the success made to date was one of the ways that they could remain supportive of the effort and not lose focus on the long-term goals. In this particular case, had we achieved only half of the benefits before losing support, the program would be considered to be unsuccessful because the long-term goal of implementing a new system with clean, authoritative data would never be accomplished. This can be a difficult process to do as each year's budget is requested and there is always need for improvement. By ensuring that accurate updates are being made, progress is in line with expectations, and goals are being achieved, the budget approval process is must more effective and easier to proceed through on an annual basis.

As has been discussed, leaders come in many forms and styles. No single trait can be identified to ensure effective leadership, but the morality and interpersonal skills of the leader have a tremendous impact on the organization, followers, and stakeholders. Therefore an examination of the individual who chooses to lead is crucial to understanding how leadership styles and traits are established, and the potential impact these can have within various organizational settings. Ultimately, anyone choosing to lead is first and foremost a person with goals, ambitions, motivators, ethics, values, and morals that drives him or her forward to achieve more.

While transformational leadership is the most common approach to achieving a high-performing productive team, situational concerns such as historical failures, upper-level management support, and demanding customers can have a negative impact on the team. Very few teams will already have reached a high-performing level on their own, and if they have, they will rarely need a new program manager to drive them forward. Most organizations try to not mess with success so will work hard to keep a leader who is achieving success in both reaching program goals and maintaining employee morale, innovation, and risk tolerance. So while you may be a transformational leader at heart, you need to look at the team and determine the most effective way to get them to be on the same page, eliminate in-fighting, and start working toward a common goal.

A single vision is one of the more crucial aspects to driving teams toward a common goal, but interpersonal skills, communication skills, positive conflict resolution, program management process, and defined roles and responsibilities are also critical. In addition, the team needs the support of the leader to overcome obstacles and to create a healthy and safe working environment for teams to move through development stages and open communication channels and trust with one another.

Support from the leader comes in many forms: dealing with customer (stakeholder) concerns, ensuring that external management does not interfere at crucial intersections, employing a process that all teams will use, and assuming responsibility for the actions of the team are all contributors to driving a team toward a high-performing work environment. In effect, a program manager must have the long-term vision clearly planted in the team members' minds, lead the team in a common direction, and ensure that there is a trustworthy and viable channel for communications.

Obstacles such as those noted previously can often be barriers to leadership. Leaders, regardless of the situation, demonstrate a set of traits that assists in overcoming barriers to success. However, even looking at leadership through various categorizations, it is still difficult to attribute leadership to a single set of traits: "Fifty years of study have failed to produce one personality trait or set of qualities that can be used to discriminate between leaders and non-leaders" (Jennings 1961, 2). Trait theory states that all effective leaders exhibit certain characteristics and traits, or a sufficient representation of characteristics that other effective leaders possess (Covey 1991; Gardner 1990; Hesslebein, Goldsmith, and Bechkard 1996; Kouzes and Posner 1995). Therefore it is the culmination of multiple traits, situational aspects, values, beliefs, ethics, and a willingness to lead that combine to form a true leader. Bass (1990) points out that "the desire to achieve, to complete tasks successfully, is a personal trait associated with those who emerge and succeed as leaders and non-leaders" (147).

These personal leadership traits are noticeable not only by the observer but also by the constituents of the team. The traits are commonly documented across various schools, methodological approaches, and belief sets. Kirkpatrick and Locke (1991) describe six specific traits common to all successful leaders—drive, desire to lead, self-confidence, honesty (integrity), cognitive ability, and industry knowledge—and those "leaders who have the requisite traits . . . have a considerable advantage over those who lack the traits." Of the six traits listed by Kirkpatrick and Locke, five can be taught or inspired, whereas one is a talent that is best achieved through intelligence, experience, and education: "Cognitive ability (not to be confused with knowledge) is probably the least trainable of the six traits" (55). However, one of the key choices that must be made by a leader is that of honesty: "Honesty does not require skill building; it is a virtue one achieves or rejects by choice" (58).

Leaders who demonstrate these traits are easy for teams to follow, while those who do not can actually be detrimental to a team's ability to accomplish its objective. Good leadership brings a team together, and bad leadership can pull the team apart potentially damaging the greater organization they reside in. "Ultimately, the negative consequences of wrong leadership choices are both expensive and well-publicized" (Ilan and Higgins 2005). Leadership that focuses on motivators outside of defined organizational goals, strategic objectives, and organizational improvement efforts can work against the grain of the firm and will not receive the support of the organization as a whole. Therefore leadership must be focused to align with strategic objectives to achieve a common goal.

The discussion on leadership, when quantified into a set of tangible traits, behavior patterns, and skills, can then continue into a conversation on born versus made leaders: "Regardless of whether leaders are born or made or some combination of both, it is unequivocally clear that leaders are not like other people. Leaders do not have to be great men or women by being intellectual geniuses or omniscient prophets to succeed, but they all have the 'right stuff' and this stuff is not equally present in all people" (Kirkpatrick and Locke 1991, 59).

Leadership can be evaluated in four categories: (1) traits and personalities, (2) behavioral theories, (3) situational contingency and cognitive models, and (4) transformational models (Chemers 2000; Fairholm 1998). These four categories break out the various aspects of leadership into a hierarchy that can be used to evaluate a leader. Another way of looking at leadership can be shown through a skills-based leadership framework: (1) skills and competencies, (2) traits and personalities, (3) motivations, values, and principles, and (4) styles and situations (Mumford et al. 2000). Either way leadership is evaluated, individual traits are vital, and situational factors must be included in the equation.

Whereas the leadership theories listed earlier and the particular traits of leaders are most applicable to executive leaders running organizations or divisions, program managers have a different challenge to face. They are often called in to solve a situation or develop a program within a culture that they have no control over but that has impacted the team in either positive or negative ways. Program managers must have the ability to recognize the situation and culture of the organization, and whether it is comfortable or not, to adopt leadership traits that will most likely benefit the program team, and to assist in achieving the benefits identified. Program managers are not responsible for changing the organizational culture, although they do often have an impact; instead they are focused on delivering the set of benefits identified in the program to the stakeholder community within the time and schedule of the program, managing cost, scope, and risk.

While a CEO's leadership style will affect an organization as a whole, and many of us envision ourselves as that Fortune 500 CEO in the future, it is vital that program managers understand the goals of the program and demonstrate the capability to leverage leadership styles and traits that are most effective for the team. This is not to say that program managers who initially establish a command-and-control structure to get the team moving in the proper direction are required to maintain that style throughout the life cycle of the program; rather, as the team grows and starts to understand the objectives of the program, their contributions, and their responsibilities, program managers should shift to a more comfortable management style for both themselves and the team.

I find that transformational leadership is my comfort zone, and while I would love every team to respond to that style, situational, cultural, and environmental factors may require me to adopt a more directive set of traits to move the program forward. However, once the team is running, conflict resolution is positive, and each team member understands his or her roles and responsibilities, I have a tendency to shift back to where my comfort zone as a leader resides, in a transformational state trusting the team, empowering innovation and creativity, and leveraging their knowledge and expertise to accomplish the goals of the program.

For example, I have had project managers assigned to a program I was responsible for who insisted that they knew the only way to manage projects. Unfortunately, we were not in alignment with the direction and governance of the program, and monitoring and controlling aspects such as earned value were not being reported. In this situation, I was forced to adopt a command-and-control structure that required my project managers to give weekly status reports with updated EVA numbers. Each project manager had to produce a common report that could be consolidated into a centralized dashboard demonstrating the program health and progress. While quite a few of the project managers felt this was unnecessary and were convinced that they could manage their projects without oversight, the program was complicated enough that visibility was required.

Over time, reporting EVA numbers and giving weekly status reports generated a set of trends that could be used by management to demonstrate how the program was progressing and how each component effort was contributing to the overall objectives. The most adamant of project managers started to see the value of the governance program and became advocates for program governance, operating as change agents for the subsequent projects.

Once this was accomplished, I was able to relax my management control to a more transformational state, supporting the project managers in their endeavors and empowering them to make decisions related to their projects. Because I had complete visibility into the project efforts, this became a much easier process, enabling me to report to stakeholders and management and allowing the project managers to focus on their project objectives.

As time progressed the project managers became more empowered, made their own decisions, contributed to program initiatives with innovative ideas, solved conflicts with their peers in a positive way, and made valuable contributions to process improvement for the program.

TRAIT THEORIES

Trait theories are based on the idea that if a leader is endowed with superior qualities that differentiate him or her from others, it should be possible to identify those qualities and measure them (Stogdill 1974). In 1990 Bass categorized leadership traits as follows:

1. Physical characteristics, such as age, height, and appearance
2. Social background characteristics, such as education, social status, and mobility
3. Intelligence and ability characteristics, such as knowledge, judgment, and fluency of speech
4. Personality characteristics, such as dominance, emotional control, and self-confidence
5. Task-related characteristics, such as drive to achieve, desire to excel, and task orientation
6. Social characteristics, such as the ability to enlist cooperation and administrative ability

Although the physical characteristics are somewhat conjectural and based on the cultural thinking at the time, sufficient evidence has been provided to support these traits. These physical, personality, social, and communication-based skills are all demonstrated in well-known leaders throughout history such as George Patton, George Washington, Mahatma Gandhi, and Barack Obama. Each demonstrated an ability to motivate and generate enthusiasm in followers and was able to achieve greatness personally and professionally.

When looking at the demonstrable attributes of a leader, it is apparent that ethics and integrity are also integral to successful leaders. Although schools try to teach ethics, morals, and values, it is questionable whether these aspects of leadership and behavior can be taught. For example Kerr (2007) raises a number of concerns about current leaders. His article on Christopher Columbus claims that the actual definition of ethical behavior must be shifted to meet the current business practices of corporations, rather than the corporations shifting their practices to the current definition of ethics.

Kerr (2007) starts his article, which was written to defend unethical behavior within current environmental factors, with a quote defending his position on unethical behavior: "Columbus was not above using devious, even deceptive, techniques to keep his crew in good spirits and devoted to the common purpose. He did not forget his crew's concern for getting home, and in good time. To be sure that he would not discourage the men, he falsified his daily journal of the voyage. In noting down his estimates of distance traveled, 'he decided to reckon less than he made, so that if the voyage were long the people would not be frightened and dismayed'"(Daniel J. Boorstin 1983, 234). In this article Kerr states, "I, and almost everyone else with my sort of responsibilities, can scarcely make it through a single workday without engaging in behaviors that violate what are said to be the basic elements of ethics and integrity" (7). This in itself questions whether ethics can be effectively taught, and it asserts that the challenge faced by a leader is to act ethically regardless of the situation and that even leaders who have demonstrated tremendous success do not have the intrinsic values that are required to achieve effective leadership.

Kouzes and Posner (2007) proposed five competencies: "challenging the process, inspiring a shared vision, enabling others to act, modeling the way, encouraging the heart" (9). Leadership models approached theory from "emotional intelligence" and offered five components: "self-awareness, self-regulation, motivation, empathy, social skills" (Topping 2002, 37). Although the wording is different, the value of social skills and enabling others to act or empathize are so similar that it is the communication approach that differs more than the actual content. In a similar fashion, both Kouzes and Posner (2007) and Goleman propose the belief that "you don't have to be the smartest to be an effective leader. Rather, emotional maturity and credibility play more important roles in how well you provide leadership inside an organization" (Topping 2002, 28). Yet at the same time Topping goes on to say, "It is commonly thought today that enlightened leaders are participative, encouraging and focused on the development of their people" (4), all factors that require a certain level of intelligence, competence, and value system. Kouzes and Posner postulate in their model that "credibility is the foundation of leadership" (4), and even Welch in his description of a vision for General Electric stated that "Speed, Simplicity and Self-Confidence" were required to move the organization forward (135).

With all this evidence pointing toward competence, values, ethics, and motivation, the theory of organizational influence on leadership ability has an even greater foundation. This further begs the question of whether someone who is not intelligent, competent, or skillful is capable of achieving a level of self-confidence that will inspire followers. Topping (2002) states, "Followers, in order to follow, need to have some confidence in the ability of their leader. Part of that feeling of confidence in their leader, comes from their leader's confidence in him or herself" (16).

Styles of leadership were described work by Lewin and associates in the late 1930s in studies conducted at Iowa State University. Social groups were experimentally controlled: some groups were allowed to make decisions through group interaction (democratic), others were told what to do and how (autocratic), and others were simply left to their own accords (laissez-faire). The end result of Lewin et al.'s work (which drove many additional research efforts) was the theory that a person's behavior was a function of the person and the environment (Lewin 1936; Lewin and Lippet 1938; Lewin, Lippet, and White 1939).

The continuum between insecurity and arrogance, with self-confidence in between, implies that those with effective skills and abilities will be able to gain a certain level of self-confidence without becoming arrogant. *Merriam Webster's Collegiate Dictionary* (2003) defines "arrogance" as "exaggerating or [being] disposed to exaggerate one's own worth or importance." In the case of an incompetent, should he or she display self-confidence without the skill sets necessary to achieve it, he or she must, by definition, cross the line into arrogance.

As a program manager, you will need to demonstrate confidence in yourself, your team, and the defined goals of the program. Many may think that you are being unrealistic or have no basis of understanding, but if you start with disbelief and a lack of confidence, the program participants will pick up on that and will believe that the program is not possible. As a leader you must demonstrate your commitment to success, looking for the resources, tools, and talents that are needed to accomplish the goals. At the same time, a leader must demonstrate a level of ethics that team members value and communication skills that allow the leader to work with both the program teams as well as the stakeholders of the effort. Through this, teams will gain confidence in the leader's ability to lead and the potential for success in the program.

I have set up data centers across the world for a single program. Initially, this looked to be an impossible effort as each center was to come on line at the same time and the centers would consist of up to 100 servers, routers, switches, failover devices, tape backup units, and cabling. To accomplish all of this we had to expand our team to include partners such as Dell, Cisco, F5, and the data center technicians. It would have been impossible for us to fly all over the world and have simultaneous startups. However, from the onset I demonstrated a strong belief that we could accomplish the goals and just needed the teams to assist in developing strategies that would make us successful. It was surprising how many of our vendors/partners were willing to join in and support our efforts, and a number of white papers were later generated on the program documenting the advantages of open and honest communication with program teams to meet their goals from the onset of the effort.

VALUES ETHICS AND BELIEFS

Values-based leadership and management by values have become popular themes in management research. Researchers such as Blanchard and O'Connor (1997), Despain and Converse (2003), and O'Toole (1996) have all contributed to a knowledge base that demonstrates that the set of values a leader holds has specific impacts on his or her ability to be successful in varied environments and with multiple challenges. A values-based leader focuses on the use of his or her personal values in decision making and on management focus within varied organizational environments and is recognized as a key tool for leaders to motivate followers.

As program managers, we cannot leave our values behind when taking on new programs. They permeate every decision that we make and contribute to how we interact with people. Understanding values-based leadership is crucial to approaching different environments, cultures, and personalities with different styles of leadership. It is vital to understand that at no time am I recommending that you do something that is against your values set. I'm simply saying that adjusting your leadership style can be more effective when facing challenges. However, stealing, bribing, or cheating as part of your values is never brought into question. We must align our values with the leadership style we undertake as well as the values of our team members so that we can be effective leaders without compromising our beliefs or putting individuals in a position where their belief set is being questioned.

Rokeach (1973) offers a definition of a value as "an enduring belief that a specific mode of conduct or end-state of existence is personally or socially preferable to an opposite or converse mode of conduct or end state existence" (5). As such, values drive behavior in individuals, but can be ambiguous and aimed at the outcome that is socially acceptable or benefits the organization. The motivational, cognitive, affective, and behavioral aspects of values can be vital in driving individual behaviors: "Instrumental values are motivating because the idealized modes of behavior they are concerned with are perceived to be instrumental to the attainment of the desired end-goals. Terminal values are motivating because they represent supergoals beyond the immediate biologically urgent goals" (Rokeach 1973, 14). Burns (1978) questions whether values are simply a motivator or whether they offer a stronger driving factor of behavior. What is your view and what does it mean for the program manager?

The two types of values, instrumental and terminal, describe short-term and long-term goals and achievements. Terminal values and end values refer to an end state of existence and are intertwined with goals, purpose, and standards of performance (Burns 1978; Rokeach 1973). As a program manager, your leadership style may adjust on an instrumental basis to meet the needs of the organization or team members. Therefore if you assume a command-and-control approach, you are not directing people to do things that are not in alignment with the program's goals and objectives. However, the terminal values refer to the outcomes of the program as a whole. Your values must be in line with the outcomes or you will ultimately be unsuccessful regardless of your leadership style.

In further evaluating the effect of values on leadership behavior and personality, it is noted that values are a part of the individual and as such are internal drivers toward a set of objectives: "Compared to cognitive certitudes . . . values are internalized so deeply they define personality and behavior as well as consciously and unconsciously held attitudes" (Rokeach 1973, 14). Therefore values that are deeply internalized are generally strong enough to influence behavior even in situations where conflicting motivations exist (Maio et al. 2001). Values can be used to elicit both benevolent and achievement motivation (Treviño, Brown, and Pincus-Hartman 2003).

The researchers listed in the preceding paragraphs agree that values are prime motivators of behaviors and that value alignment influences the behavior of groups and teams. Values are seen as "relatively enduring constructs that describe characteristics of individuals as well as organizations" (Meglino, Ravlin, and Adkins 1992, 17).

Prilletensky (2000) sets out three sets of values to guide individual and organizational behavior: (1) values for personal wellness, (2) values for collective wellness, and (3) values for relational wellness, where wellness is defined as "a satisfactory state of affairs brought about by the fulfillment of basic needs" (000).

Enderle (1987) described managerial ethical leadership as "applied ethics," which at first appears more related to transactional leadership than transformational leadership. He described three interconnected, normative-ethical tasks of leadership: (1) perceiving, interpreting, and creating reality, (2) being responsible for the effects of one's decisions on others, and (3) being responsible for implantation of the company goals. You need a transition.

Ethical leadership is about behavior and is visible in leaders who (1) create and institutionalize, (2) stick to principles and standards, (3) are uncompromising in the practice of value-based management, (4) do not tolerate ethical lapses, (5) use rewards and punishment to hold people accountable to standards, and (6) are concerned about the interests of multiple stakeholders and serving the greater good (Treviño, Brown, and Pincus-Hartman 2003). To be an ethical leader "the executive must engage in behaviors that are socially salient, making the executive stand out as an ethical figure against an ethically neutral ground" (Treviño, Brown, and Pincus-Hartman 2003).

Ethically neutral leaders can be successful but generally (1) can be more self-centered and interested in personal gain, (2) have a short-term focus on the bottom line, (3) use power in a negative way, (4) are less aware of ethical issues, and (5) are less concerned about leaving the world a better place for the future (Treviño, Brown, and Pincus-Hartman 2003). Transactional leadership adds the mutual or utilitarian motive to behaviors within these tasks, where the ends may justify the means. The goals are pragmatic, and there is an independent, individualistic, self-centric leadership style (Kanungo 2001). Conversely, Kanungo identified the moral altruism of transformational leadership, where means justify the end, goals are altruistic, and there is an independent, collective, socio-centric leadership style. [You need a 'so what' here to maintain reader's interest.]

There is a distinction between values that are consciously shared and lived and those that remain unconscious and not discussed. Espoused values may be known, because for instance they are posted on the walls of offices or the pages of a website, but they may not be shared, taken for granted, negotiated, or even discussed, and most likely may not be remembered (Barrett 1998). The espoused values of the facilitating idealist would be well known, more than just words on the wall, and supported by underlying, taken-for-granted values created by the ethical leader as part of the organizational culture (Schein 2004).

To summarize the research on values-based leadership, for a very short period of time I went to work as a program manager for a "market research firm." Unfortunately, in my first week what I discovered was that the organization was collecting data through malware and proxy intervention including personally identifying information such as Social Security numbers, credit card numbers, phone numbers, and birth dates. My values were at odds with the intention of the business, and I was hard-pressed to help them be more successful.

Thankfully, my program was to modify an external website to facilitate multinational languages into it outside of the malware and spyware that the organization collected. As a program manager, I adopted a management style that allowed me to address concerns about the IT staff, drive projects to completion, and reach a point where a single website was serving up information in over seventeen languages. Very soon after this program reached a closing point, I located another position, enabling me to operate with a set of values and ethics that I believed in. Had I attempted to stay or to work on the spyware the firm was deploying, I would have had an internal struggle between the need to do an effective job as a program manager and contributing to a problem that we see way too often on the Internet, the collection and sharing of personally identifying information.

10

Leadership in Program Management

It is crucial to understand that the leadership style for program management will quite often differ from the style a CEO or project manager may use. A program manager is responsible for a long-term vision leveraging many projects that together achieve a set of benefits not possible from a single project. Therefore the program manager must have plans for starting new projects, transitioning projects to operational teams, managing a longer-term budget, and accepting the responsibility of communicating with all stakeholders from every project as well as those specifically focused on the program. Additionally, the program manager must constantly ensure that the benefits of the program are understood by and communicated to all, and identify when these benefits are achieved. And whereas a project will often last only six to nine months, a program may very well be a multiyear event that has an ongoing impact on stakeholders with intermittent benefits being realized.

One of my favorite programs was the conversion of a very busy road into a freeway. The program manager assigned had over fifteen intersections that needed to be converted from stoplights to overpasses. The long-term benefits were simple: traffic would move faster and decrease commuting time for the average person, and speeds would be raised from thirty-five miles per hour to fifty-five miles per hour as the road was converted to a freeway.

The stakeholder community for this program was the millions who used this road every day, the government agencies who were paying for the effort, and the contractors hired to build the overpasses. The budget was established based on both the cost of the overpasses and the cost of communicating with such a wide base of stakeholders.

Initially, the program manager built a website to communicate the long-term vision, demonstrate the current delays in hours/minutes to traverse the road at various times, and display the schedule for conversion of intersections. In this website he warned that the impact to the average commuter would be kept to a minimum as he had requested all bridge work to be done in the evenings, but he also consistently reminded the commuters of the long-term advantages and schedules for each intersection.

As the program progressed, the website was consistently updated with progress as well as any delays encountered and the strategies in play to overcome the delays. The road was depicted in red, yellow, and green based on the traffic patterns, and through the real-time use of cameras was accurately depicting travel time. As each intersection was completed, the website was updated with celebratory information, a ribbon-cutting event took place, local newspapers were informed, and town hall meetings were held.

What should have been a truly painful program with constant complaints by average commuters had turned into a very successful effort that maintained ongoing communication and celebrated success. At the completion of the effort, the cost overruns were minimal because the program had been so successful with the stakeholder community that it overwhelmed the government with letters of congratulations and positive feedback. Imagine celebrating road construction—not a very common outcome.

CASE STUDY

You have just been assigned as a program manager for a new privately owned cellular firm. Your program will affect many, if not all, aspects of the firm and will require interaction from all departments such as finance, legal, human resources, investments, sales, and information technology. The program will be to develop, test, market, and sell a new product for the organization.

Although you interviewed with the CEO, CFO, COO, CTO, and board of directors, you have not yet met with the individual project managers or the teams that will work on the delivery of the new product. The CEO informs you that you have absolute discretion over the team and that he personally knew of a number of resources that had outlived their usefulness at the firm. He grants you the right to terminate staff on an as needed basis. In addition, he expresses his desire to retire from the firm once they have become profitable again.

The CFO explains that the budget has to be constrained because the firm has been rapidly losing market share and is heading toward becoming obsolete in the market. This new product is their only hope for pulling out of the dive that the company is in.

The CTO explains to you that they have a huge technology deficit and that the teams are not trained in the latest technology or potential new technologies, yet there are no training dollars assigned to the program.

Your first step in the program is to sit down with the project managers assigned and you immediately determine that there is no understanding of the PMI or its best practices. The teams have all been acting as unique groups, and while some have been successful, most have failed at their efforts in terms of either cost, quality or schedule. The team is very negative and spends more time discussing the organizational issues and less time on the projects that they have managed in the past.

As you walk through the office, the team members are sitting in their cubes, and while they look up at you, you notice no smiles or interactions between individuals. Each person sits in front of his or her computer and there is dead silence on the floor.

CASE STUDY QUIZ

1. What organizational issues can you identify from the case study?
2. Based on the information provided, is the organization's risk tolerance high or low?
3. How would you define the communication style of the project managers and individual team members?
4. What would your initial days look like in the creation of this program?
5. What leadership techniques will you employ and in what order? How will you know when to transition from one to the next?

DISCUSSION QUESTIONS

The organization just described is obviously a very toxic one, where failure has become commonplace and even the CEO is looking to abandon the firm at the first opportunity. Based on the information provided, how would you describe the morale of the teams and what level of empowerment would you expect to see from them?

In the case study, the project managers have not had any training in project management methodologies and are not aware of the benefits that can be achieved by implementing best practices. How can you as the program manager achieve your goals of (1) strategy alignment, (2) program benefits management, (3) program stakeholder engagement, (4) program governance, and (5) program life cycle management? What leadership styles would you employ and what would you see as transition points if multiple styles are implemented?

Section III

Leadership and Teams

To lead people, walk beside them . . . As for the best leaders, the people do not notice their existence. The next best, the people honor and praise. The next, the people fear; and the next, the people hate . . . When the best leader's work is done the people say, "We did it ourselves!"

—Lao-tzu

11

Building Teams

> Never tell people how to do things. Tell them what to do and they will surprise you with their ingenuity.
>
> **—General George Patton**

Teams are not simply a collection of people thrown together capable of establishing common objectives. Instead, every person on a team will bring with them political opinions, personality traits, unique skill sets, communication abilities, and personal agendas. In some cases the very concept of being added to a team may generate resentment if the individual has become accustomed to working alone or is pulled from working with a manager that he or she has come to respect and like. Should that manager have been removed from the organization, the resentment can even be very hostile. As a program manager, the task at hand is to take these very different people and personalities and blend them into an effective team, building on their strengths and working through individual weaknesses.

In looking at team development, it is critical that a program manager first understand some basics about teams. Every team, no matter the circumstances around the program, organization, past experience, or the environment, will go through a very tangible set of steps. These steps are more based on who we are as people than on the objectives of the program or the organizational environment. The five stages of team development best describe the process of bringing people together into a cohesive whole and can be recognized for what they are by a successful program manager. There is no defined set of time in which a team will progress from one stage to another, but the program manager can definitely help the team in the progression. Some teams never reach the highest levels, producing at less than effective performance levels; whereas others quickly move through the stages, optimizing performance and communication to such

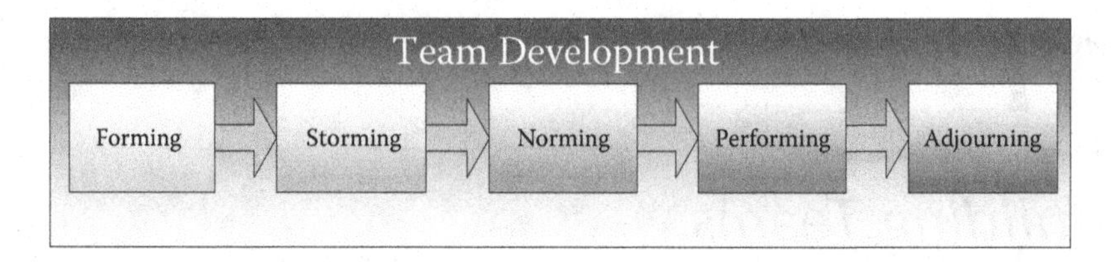

FIGURE 11.1
Team development.

a point that it is a sad and lonely time when the team completes its work and moves on to other tasks.

Bruce Tuchman (1965) provided five stages of team development:

1. Forming
2. Storming
3. Norming
4. Performing
5. Adjourning or mourning (384)

But the speed and effectiveness at which the team advances through these stages are generally set by the leader. While the five stages are vital for teams to begin to be effective, the length of time spent in the early stages can be manipulated and driven by the program manager to push toward performing, where the real work is accomplished. The leader must know not only where the team is at currently in the process but also how to push through to the next stage as quickly as possible.

If you have not come across the five stages of team development, they are described in greater detail in Figure 11.1 and the sections below. Every team that I have worked has gone through this process, whether quickly or very slowly, and it is only when a team achieves the performing stage will they be able to achieve a high-performing level. The five stages can be described as follows:

Forming – The team meets for the first time as a single unit and there are generally smiles all around. Each member is operating on his or her best behavior and, as a general rule, assumes that the other team members all have a similar background and interest. In forming, roles and responsibilities have not yet been established. The senior members assume that they will play a senior role and that there may be others more or less suited for the role they see themselves in.

Storming – At this point, the team has started work and conflict breaks out. Team members "jockey" for position, conflict with each other, and attempt to position themselves in various roles. The conflict can be either negative or positive but needs to be effectively managed. At this point in the team development phase, they will be the least productive and a program manager will have his or her hands full working to overcome conflict in positive ways.

Norming – In the norming stage, conflict lessens and team members start to better understand what their roles and responsibilities are. They begin to focus on the work and not the relationships and spend less time positioning themselves within the team. Oftentimes, this is where team development will stop without good leadership. The work is getting done, but not at the most efficient or effective rate.

Performing – Performing is the ideal state for a team. There are times when the team does not make it to this stage, but an HPT will, and it is here that both the trust between the team members and productivity improve dramatically. The team starts to trust one another, working toward a common goal, handling conflicts internally in positive ways, and operating at maximum capacity. Problems are handled quickly and effectively, and risks are recognized and resolved per initial planning or even new approaches.

Adjourning (Mourning) – This is the stage when the effort has been completed and teams will be dispersed to work on other projects/programs. Although it is the adjourning stage, it is also referred to as the mourning stage as people have formed emotional bonds and friendships and dispersing the team is a sad occasion.

Each of these stages helps to bring people together and to achieve a more positive working relationship. If a team is not progressing effectively through the stages, the leader must identify the reasons, develop a strategy, and implement it to push them forward. A leader who does not work with conflict resolution or has a poor vision will curtail the growth of the team, who then may end up being stuck at a storming or norming stage and will not be able to move through to the performing stage where the work is done effectively and efficiently.

It is also the leader who can push the team through the storming stage faster by teaching and demonstrating positive conflict management, promoting open and honest communication, and helping the team to develop

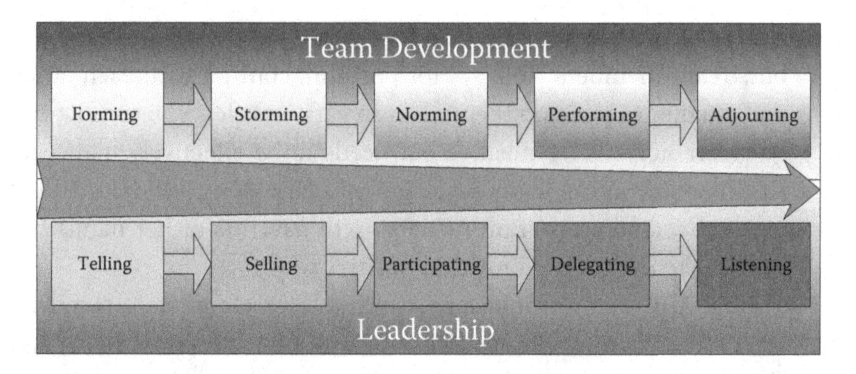

FIGURE 11.2
Team development and leadership.

patterns that eliminate hostile confrontation and promote innovation and positive conversation in an effective way.

As illustrated in Figure 11.2, effective leaders will drive the team development through a process of telling (informing), selling (gaining consensus), participating (negotiating roles, responsibilities, and tasks), delegating (empowering staff), and listening (lessons learned, process improvement). Through these processes the leader facilitates team development, empowers staff, and assists in the transition between phases. Each stage of team development has a corresponding responsibility area for the leader that drives the team to push to the next stage. Without this leadership involvement, teams can become stuck in one phase and not able to successfully move to the next. The goal is to bring the team to the performing stage as quickly as possible as that stage is where the greatest productivity is achieved.

During the storming phase, leaders can leverage a responsibility assignment matrix (RAM) or RACI chart (Responsible, Accountable, Consulted, Informed). This tool enables a leader to define individual roles and responsibilities and can assist in driving a team through the storming stage by clearly identifying who will be responsible for what aspects of the program. RACI enables a program manager to communicate expectations to team members and helps to decrease the posturing and assumptions made by individual team members.

A program manager will have multiple project teams and will work most closely with the individual project managers to create effective teams. Project managers will most likely come to the table with a set of previous team management experiences and an understanding of the project management process. As a general rule, project managers will have a better understanding of team development and will be easier to work with in pushing through the storming phase. On the other hand, project managers will have team members that may or may not have worked on effective teams in the past. These team members may be recognizing the team development stages for the first time and may need to be trained in effective communication and conflict resolution techniques to help them in working with their peers.

Teams can also benefit through either formal or informal team building. The intent is to bring out the best in a team to ensure self-development, positive communication, leadership, and the ability to work closely together as a single unit to solve problems. Team building can be a wide range of activities from very formal multiday retreats focused on team challenges to a much more informal approach of bonding, simulations, problem solving, or even communication building. One of the more effective methods I have employed was a very simple lunch held on a regular basis so that teams could get out of the work environment and actually spend a few minutes getting to know each other. My only requirements for this lunch were that each person choose a different seat at each lunch so he or she could meet and get to know new people, and that work not be discussed during the lunch. Through this process, team members began to gain an understanding of who they were working for and to form friendships and respect for one another.

Through either formal or informal team-building activities, team dynamics will start to develop, enabling individuals to build trust in each other, share knowledge, problem solve, and handle positive conflict resolution. We often look to team building to overcome individual or group conflicts so that they can be brought into the open and resolved together.

The focus of team development is to move from a directive approach to a more supportive approach enabling the team to drive their own performance. In team development the leader will assign a goal for the team to accomplish. The goal can be a simple or complex task such as building a fence or crossing a river together without any team member touching the water.

As a team undertakes the task, the leader will assist in coaching the team toward potential solutions that could be leveraged and supporting the efforts of the team to achieve the desired outcome. As the team becomes more self-sufficient and forms their own approaches for solving problems, the leader begins to delegate responsibility and allows the team to determine their own direction for problem solving. Over time, the

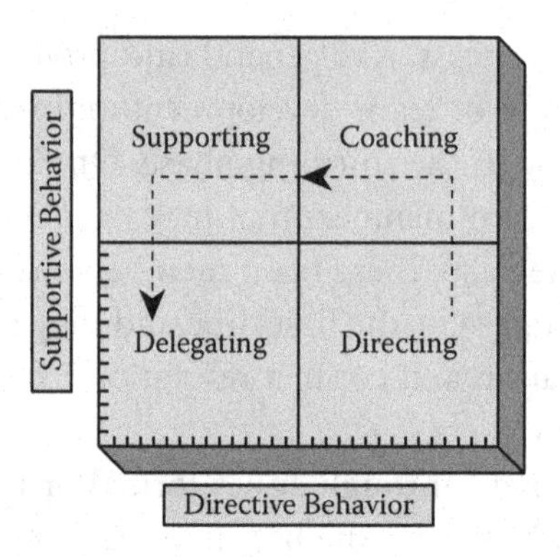

FIGURE 11.3
Leadership approaches for team development.

leader is more supportive of the team and provides less direction, enabling the team to develop solutions leveraging their own expertise and teamwork (see Figure 11.3).

An HPT needs to have specific and meaningful performance goals for each member and a clear and concise vision for the effort. It must actively work with each team member to ensure that the work components are being done well and on time in alignment with the vision. The leader must ensure that effective and efficient approaches are being followed. And the team members need to believe in and support the vision, assuming responsibility for their contributions.

In an HPT, there are more advanced qualities that teams need to have in a general sense. But the unique quality in an HPT is that the team members are intrinsically motivated, self-managed, have an inner need and ambition to go after bigger challenges, and demonstrate a work ethic that creates a deeper commitment to the collective mission and vision.

Teams respond to enthusiastic, confident, personable, and quality leaders to build positive working environments. Authoritative managers, people who speak and expect that no one will question their directives, can force an unnecessarily formalized and hostile environment that creates a more uncomfortable work space and inadvertently decreases communication, innovation, creativity, and personal investment. Such an environment will create a culture where minor mistakes are seen as major and where individuals have a tendency to withhold information and work product until they have achieved a level of perfection. A less formal workplace will have draft versions shared earlier and often avoid rework because of greater flexibility and more open communication with team members and the leadership teams. According to the book *The Wisdom of Teams* (Katzenback and Smith 1993), these strong extensions grow out of an intense commitment to the team's mutual purpose.

Work approaches are another determinant of team performance. Work approaches comprise a whole host of team work processes such as:

- Decision making norms within the team
- Positive conflict resolution processes
- Innovative approaches to problem solving
- Leveraging of best practices and standards matching industry norms
- Communication management to ensure that meetings and informal conversations are efficient and effective
- Developing informal processes for hand-offs and task completion to enable all team members an opportunity to achieve success

When a team is able to successfully integrate various aspects of healthy and proactive functioning, it paves the way for a far better team performance than a team that struggles to find mutually acceptable methods to move the work forward. For example, better work approaches can ensure better planning and scheduling of activities, quicker decisions, rapid response to customers, meeting deadlines, and so on. There is no need for creating new standards and processes if they are available through industry best practices. Instead, leveraging these standards allows team members to have a foundation on which all can interact, enabling the individuals to have expectations of other team members in work product, hand-offs, transition points, and status reporting. Methodologies such as PMI's *PMBOK*, programming guidelines, contractual obligations, and technology road maps all enable teams to establish a clear foundation, eliminating the need to create new standards, and deal with dissension when those standards are not clear or effective.

High performance teams achieve mutual accountability. Mutual accountability is the collective responsibility of the team toward generating results and achieving success. Mutual accountability implies an implicit acknowledgment of the joint accountability of all team members toward a common purpose, in addition to the individual's obligations in his or her specific roles. This creates a supportive environment within the team, and the performance of the team improves in the presence of this type of mutual support and cohesion.

Complementary skills are a necessity in most teams. Most team tasks require multiple skills, and when the team members have complementary skills that are well balanced and congruent to the task, team performance is bound to be raised. Interchangeable skills can be an asset in some businesses, because the team members can depend on one another to jointly accomplish a task.

To sum up, the qualities that seem to foster high team performance are primarily a cut above those of an average team. It is certainly not easy to create an HPT with all these qualities, but an organization can provide the building blocks with a few necessary measures such as the following:

- Implement challenging and inspiring performance goals that facilitate the feeling of achievement.
- Empower team members to achieve individual success and facilitate leadership abilities where possible.
- Provide effective training and reference materials to team members to assist in their personal growth and increase expertise levels.
- When possible, bring teams together with complementary skills to form a foundation by which the team can rely on each other for task completion.

12

Team Dynamics

> Leadership is not a magnetic personality—that can just as well be a glib tongue. It is not "making friends and influencing people"—that is flattery. Leadership is lifting a person's vision to high sights, the raising of a person's performance to a higher standard, the building of a personality beyond its normal limitations.
>
> **—Peter F. Drucker**

Program managers are faced with unique challenges relating to team dynamics and development. In many organizations teams can be assigned on a temporary basis (functional) or a full-time basis (projectized). Teams that are assigned on a part-time basis still have other responsibilities and are not 100 percent focused on the outcome of the program. These team members will have trouble focusing on the tasks assigned and will often be called on to perform tasks outside of the effort, leaving the tasks in limbo. To manage these individuals effectively, backup strategies and open communication are necessary to ensure that when they are tasked with efforts outside of the program, they can easily communicate their conflict and other team members can step up to complete unfinished tasks.

In addition, as a program manager you are responsible for your program team and the project managers assigned to you. Each project manager is then responsible for his or her individual team members, and each project manager will have his or her own leadership style and approach to build a team and to lead projects. The program manager will work with individual project managers to overcome obstacles and provide management strategies to assist in successful project completion. In some cases, a program manager can leverage resources across multiple projects and reassign resources for support from one project to another. As a key success factor, the program manager sees the overall program team and will be aware of resources not fully allocated, while a project manager will see only the project resources assigned. This provides the program manager with a higher level view and additional tools to enable project success in the event of a slowdown or a functional resource being tasked outside of the project. Yet it is only through open and honest communication between team members and program managers that the need for additional resources can be established. If a project manager feels that communicating needs would be an indicator of failure or that his or her request would not be seen in a positive light, the request may never be made, leaving the project (and therefore the program) at risk.

Program managers must be able to develop teams and support project managers in such a way that the overall program is working as a cohesive organization without directly undermining the leadership of the project manager. Because so many of us come from a project management background, this can be one of the hardest challenges to face. As a program manager, you must be able to decide when to allow the project manager to guide his or her teams and when to step in and assist the project manager in resolving issues and building teams to a more effective level.

My goal is to build an HPT regardless of the organization, culture, environment, or past experiences that the team members have faced. This is a daunting challenge as many have been beaten down, or taken the blame for project and programs that have failed in the past, regardless of their ability to resolve the issue. To make a team successful the span of control (number of members) must be small enough so that they can be brought together easily and communicate with each other on a regular basis. The program manager is responsible for communicating the vision in a clear and effective manner so that the team can better understand both the goals of the effort as well as its contribution to the organizational objectives.

It is often difficult to not walk into a program with guns blazing and take over every aspect of the program. Because we know where we come from and what our experience level is, we must temper that and spend time observing how things have been done and how teams are working together. When starting a new program or stepping into an existing one, spend some time looking around at the project managers and team members to observe their communication skills, their collaboration styles, their understanding of roles and responsibilities, techniques for resolving conflicts and their responses to the program objectives. A successful team is one that displays open communication, positive conflict resolution, and collaboration; has the proper skill sets and training; and respects the role

TABLE 12.1

Analyzing Program Teams

Role	Goal
Problem solving	Does the team have effective communication and conflict resolution? Are they empowered to find solutions?
Communication	Is the interpersonal communication between team members positive, enabling open communication? Are there individuals who struggle with participating in team discussions?
Interpersonal relationships	Do team members have past experiences with each other that have been unresolved? Are there members who conflict? Is there a strategy for overcoming these conflicts and moving forward in a more healthy relationship?
Conflict resolution	Does the team operate with positive conflict resolution or negative? Are there personal issues that affect interaction?
Skill sets	Do team members have proper skill sets for the tasks at hand? Have they been provided resources and training to maintain their technical knowledge?
Motivation	What motivates the various players on the team? Is it the success of the project or personal success?
Leadership	How does the team react to leadership? What forms of leadership have been effective in the past? What areas of concern should be identified?
Organization	How does the organization treat employees? What forms of recognition or rewards are in place? Is the organization a positive environment or are there negative (toxic) aspects that decrease morale within individuals?

that leadership brings to the program. As such, the most effective way to start with a new team is to run through a formal evaluation of the team's interaction, skill sets, experience, relationships, and personal drivers to better understand who they are and what motivates them. A program manager should start with a formal process for analyzing the team members and environment (see Table 12.1).

I once had the opportunity to work for a financial organization that had been in existence for about thirty years. The organization had progressed to dealing with only very high dollar clients, but had become a technology-deficient firm demanding extensive manual labor to accomplish simple tasks. The IT team was incredibly small and had been through a serious of very limited managers ranging from command-and-control to transactional. To achieve the strategic objectives of the organization, the IT team required a truly transformational leader capable of driving organizational change. The challenge was less process and more performance. Teams were held responsible for the failures of outdated technology and were being measured on factors such as attendance, hourly rates, and client demands that were only a small percentage of their area of responsibility, leaving executives with an unclear understanding of how the organizational objectives were not being met.

The organization also required a more effective leader as the teams had grown tired of being faced with difficult challenges without sufficient tools or success factors and were frustrated with being measured on factors that they were not in control of. Recognizing the environmental and situational factors affecting the team, the first approach was to set up project management plans that outlined specific tasks for every project and rolled them up to the program level. Project managers involved every one of their team members in the creation of a work breakdown structure and then prepared schedules, assigned resources, and eliminated redundancies within the project. This input provided much more effective tools for the program to assess risks, identify opportunities and redundancies, and balance out key personnel. As a team we established some workable schedules that everyone felt were achievable.

However, each of the projects we were working on started to run behind schedule and required additional time and energy to be invested to maintain the schedule. At one point we had teams working eighty to one hundred hours a week on a regular basis just to achieve the work packages assigned to them. Programmers were frustrated, project managers kept pushing, and clients were getting angry with the delays, cost overruns, and quality concerns.

From a leadership perspective, I was concerned that the team may not be up to the challenges, training might be needed, or perhaps the motivation to achieve success was not focused correctly. However, I had a team that was volunteering regularly to work the hours necessary to stay on track with their schedules. So the challenge was a daunting one. We had followed both the project management processes as well as program management standards and yet were simply not making progress.

To gain a better understanding of time spent, I implemented a performance measurement program to measure the time spent on assigned tasks, any unassigned project tasks, client management times, and timesheets to compare against the times each team member was working. Though EVA was implemented, it did not provide the information that we were looking for; we had to adapt a more unique set of measures to gather the data necessary to identify the problems. EVA demonstrated that we were over budget and behind schedule, implying that the work was not being done in an effective way or that we had potential problems with staff skill sets. However, empirical evidence showed us that this was not an accurate reflection of the issues.

As a leader, I had to cajole, convince, motivate, and at times command that the additional performance measurement tools be implemented to drill into the problems that we were facing. I made a personal promise to the project managers that if we were not able to better understand what was going on, we would either eliminate the performance measurement program or dramatically modify it.

Surprisingly, after three weeks of measuring tasks one factor became abundantly clear: programmers were working at least 40 to 50 percent of their time supporting legacy-based applications and consistently received calls from favored clients and executives who asked for "quick fixes" to production-based applications. Because of environmental factors and organizational history, none of this time was logged and no formal requests were made with management. Instead, stakeholders would reach out to programmers with one-off requests and issues. With these numbers in black and white, it was blatantly obvious that within a single eight-hour day, only four to five hours were spent on project tasks while all of the undocumented tasks were taking five to six hours a day. It was no wonder that these people were consistently working overtime to simply maintain schedules and could never quite get ahead.

Initially, we resolved the issue by changing their available time from eight to four hours a day on each project and extending the delivery dates. This, of course, was not an easy sell to clients but was much more consistent with the capacity of the team. Shortly after that, we moved some programmers into a legacy support function and hired others specifically to support legacy and production systems, keeping our very best working on the new development efforts. Suddenly, we had two teams working eight hours a day without overtime. Not only did the productivity increase, our quality and deliverable schedules improved so dramatically that other teams started following our example.

Finally, team morale, communication, trust, and conflict resolution started to improve dramatically. Time was made to discuss issues encountered and to create a knowledge base repository, and quality was included at the beginning of each project initiative instead of at the end of the development life cycle.

As projects were completed on time, exceeding client expectations and maintaining a tremendously high quality standard, we started to celebrate successes, recognize achievements, and reward stars on the teams. Instead of performance measurements being an additive task, the teams started to volunteer metrics to support their actual workload and to facilitate further improvements. Most importantly, morale improved to a point where we suffered 0 percent attrition over a four-year period, down from a 30 percent rate in the previous four years.

While the team had become a successful one, other teams were not nearly as effective. Very quickly animosity grew between the different groups, and instead of sharing knowledge and tools, it became a very competitive and somewhat hostile environment. The other team resented the progress that was being made and felt that an unfair advantage had been given to the others making it impossible to compete.

Before this became an untenable situation, I had to reach out to the other division directors and start sharing knowledge, tools, and team members to assist in making both groups successful. While some of our methods were initially hard to accept, the support we received from my counterparts soon overcame change resisters, and we were able to build a much more successful organization than what had previously been considered possible.

One of the key success factors in producing successful projects and programs is partnering with disparate organizations to achieve strategic objectives. From leveraging the knowledge of an SME group to gaining insight from external subcontractors to interfacing with a network services team that provided infrastructure support for software development efforts, programs and projects would not achieve the same level of success without effective partnering.

In both a staff position and a contracting role, partnering and communication are crucial and require conscious thought and planning. Partners can often be found within the client organization—external entities as well as subcontractors. Effective partnering enables better communication; sharing of crucial historical experiences; shared knowledge of technology, risk, and change tolerance; and often gaining a greater understanding of organizational objectives.

At one client I worked with, the customer service, technical support, and design groups were all considered disparate divisions that were not included in the program teams. Achieving successful completion required reaching out to the various division directors, establishing greater communication, and requesting resources from each team participating on programs as SMEs so that programs could avoid having to rework outcomes when the services reached the marketplace. In every conversation and every request for participation, division directors were extremely positive and indicated they had been waiting for an invitation to participate, stating that they had been left out of the process for so long that they assumed they were not needed.

Involving these members helped the program and each component project to have a much greater understanding of the market and created greater value for deliverables. In addition, when a service was considered, the SMEs were able to outline how the it would be used in the marketplace, enabling the team to clarify scope at a more realistic level and avoiding providing expanded services that would not be used or were not effective to the user community.

13

High-Performing Teams (HPTs)

Lead and inspire people. Don't try to manage and manipulate people. Inventories can be managed but people must be led.

—Ross Perot

DEFINITION OF A HIGH-PERFORMING TEAM

An HPT is a group of people who work closely together, trust each other, travel toward a common vision, and have a clear path to escalate issues and gain additional information when needed. The team can make decisions quickly and resolve conflict in positive ways, working toward a common goal and understanding their personal and the overall contribution to the effort. In addition, members of an HPT are willing to put aside their personal agendas for the achievement of the program goals through cooperative efforts and open communication.

Teams can be amazingly successful when motivated, empowered, and able to work closely together; they can accomplish what seem to be miracles, and *even like each other when they're done.* In every almost every HPT, there is a very strong leader who is driving the team forward, setting vision, resolving negative conflict, and removing obstacles from the path of success. A leader such as this is someone who has established and communicated a common vision and a defined set of objectives for the program to achieve, ensuring that each team member understands the value of his or her personal contribution and trusts the leader to provide support, establish priorities, and manage stakeholders. The team members build trust in each other, and that trust ensures success for the effort and empowers the team to achieve the established objectives. While it is not always the program manager who drives trust for the team, it is one of the crucial roles that the program manager is responsible for.

HPTs "click" with each other, focus on the goals of the project and program, and are self-sustaining and self-managed. In fact, an HPT often takes on a life of its own, driving for success without management intervention, and it is all related to the leadership style imposed. Don't confuse team leadership with team building; while team building has its place and can help a team moving from the storming and norming phases, it is not the sole driver of a HPT. Instead, the leader has to establish communication, build confidence, gain acceptance of the central vision/goals, assist in establishing team skill sets, and resolve conflicts in a positive and effective manner. The leader empowers the team members to work closely together and to make strategic decisions that will overcome obstacles internally without requiring management intervention. A truly effective HPT will be able to drive themselves forward and make decisions, reporting only the results to management.

As a program manager, the focus is always on making every team an HPT. While not always possible due to relationship issues, environmental factors, and experience levels, each team needs to establish a level of trust that that allows them to work through problems without finger pointing, job loss, morale issues, or disciplinary actions. Although the scenarios that work against an HPT are endless, the ability of a leader to empower and drive the team toward success is a crucial aspect in achieving the goal.

At times, a team can believe that they have already built so many similar products (i.e., software, website, etc.) that it can be difficult to recognize that they are creating a unique product or service and therefore will face unknown challenges. Their confidence can undermine success by avoiding proper risk management approaches and not investigating various options such as make versus buy and the technology approach to be undertaken. Teams such as this need to be walked through the requirements, unique factors of the program, and long-term benefits. If the program were simply a repeat of work that had been done in the past, that work would be reused or tailored to meet the needs of the program. Instead, a program manager must work with each team to define the unique qualities of the program, expectations of stakeholders, and long-term objectives of the program.

Regardless of the effort to be undertaken, it is vital to establish leadership that the team can rely on and follow. A leader must empower his or her people to work and make decisions, keep outside influence to a minimum, and pave the road to success by overcoming obstacles and ensuring that the required resources and tools are available. Where necessary, the leader will have to ensure that proper training and reference materials are available so that team members can advance in their knowledge base and contribute the new knowledge to the program efforts. In addition, the leader must take responsibility for the team's overcoming outside obstacles that can hinder project and program success. It helps the team to function at its highest level knowing that as their program manager a good part of their position is to protect them from outside influence so they can be successful, and that they will work together to achieve success wherever humanly possible.

As previously mentioned, HPTs are often amazingly successful in the work that they undertake. That is not to say that they can't fail; but should failure occur, the team will be able to communicate it, understand the reasons for it, and use the failure as a learning experience for future efforts. However, failure is under the leader's watch and responsibility and not individuals of the team; therefore "success is the team's, failure is the manager's, and failure is not an option." A leader who brings this motto to the table will be able to bring teams together who will understand that they are the reason for success and that the leader will take responsibility if problems occur.

You might be even able to remember a team that was so much fun to work with that you enjoyed the work, the people, and the challenges. A team like this will have a leader who brought everyone together in such a way that friendships were built and the work was fun. Even if it was truly challenging, the team worked together efficiently and effectively and generally exceeded expectations and objectives.

But teams can also suffer ongoing negative conflict to the point that nothing gets done, and when the team disbands, there is hostility toward teammates, the project manager, the program manager, and the program stakeholders. Individual projects may be considered successful, but the program as a whole has failed to achieve the benefits defined and quite often gets noticed by the watchdog groups as yet another multimillion-dollar program gone bad. Conflict resolution is crucial to building HPTs and is required for teams to develop innovative thought processes, eliminate negative working environments, and encourage communication between all team members. Often, the job of positive conflict resolution and avoiding negative feelings of team members falls on the shoulders of the program manager. Conflict resolution focuses on taking negative, destructive, and abusive conflicts to a more positive collaboration of individuals' thoughts and beliefs (see Chapter 15 on Conflict Resolution).

Why some teams can hit that high-performing mark while others simply fail to achieve their goals is a key focus of this book. There are, of course, myriad reasons that programs fail, some of the most common of which are listed below:

1. Waning executive sponsorship
2. Inefficient communication among project managers
3. Scope creep (common to all programs) that is left unmanaged
4. Unexpected cost overruns
5. Poor communication
6. Bad or ineffective leaders
7. Technology changes
8. Key management changes
9. Attrition—losing a key player, especially a project manager or program manager
10. Not effectively managing stakeholder involvement and expectations

Let's take a moment and look at these ten factors that can cause a program to fail. Each of the ten can be handled through strategic planning:

Waning executive leadership – There have been a number of programs that I have undertaken where the number one risk to the program was that we lose executive support. To overcome this obstacle, two approaches can be very effective. First, have management and the program sponsor sign a letter of commitment indicating their understanding of the duration of the program, the estimated costs, and the defined benefits. As the focus moves to other initiatives, remind the executive leadership team of the importance of the program and leverage the letter of commitment signed. Second, regularly post the fact that if executive leadership wanes or shifts to new priorities, the program will be at tremendous risk. This is usually put at the top of the risk register and discussed at every status meeting along with the dollars and time spent to date. Most often leaders are not willing to throw away tangible amounts of dollars or tangible progress if they are aware that decisions made will directly affect this.

Inefficient communication between project managers – This is actually a more common issue than you might think. Project managers, by definition, focus on their project goals and can become very myopic in their focus. Sharing knowledge, resources, or risks will go against the grain of many. To overcome this hurdle communication must be built up through proper conflict resolution and direct reminders of the focus of the program, and at times the program manager may have to make authoritative decisions regarding resources and risks.

Working with project managers requires the program manager to directly understand their concerns and to help them focus on success regardless of the needs of other programs. Quite often, an SME is assigned to a project but is needed for a short duration on another effort. Because the project manager is committed to delivering his or her project on time and on budget, releasing this SME can be painful. The program manager may need to intervene to set a defined duration of use on another project that does not conflict with the critical path of the ongoing project. When the resource is released from his or her second initiative and put back on the primary project, the project manager will gain greater respect for the program and be willing to share when necessary.

Finally, hold regular (weekly or biweekly) meetings with project managers with each sharing their risk register, project plan, issues, and status—not only so the program manager has a clear understanding of the projects within the program but also so that all project managers can identify and offer potential solutions and work-arounds to assist in solving problems on all projects and the overall program.

Just as any team requires building trust and communication with one another, project managers also need to go through the five phases of team development to begin to work toward an HPT.

Unmanaged scope creep – All projects, programs, and operational initiatives are faced with potential scope creep to the effort. From a program perspective, we define a program charter, program management plan, and change management plan to handle issues that arise during the execution of a program. Projects will also build equivalent planning documentation but will reference the program change management plan if a sponsor or stakeholder is insistent that the product be modified. Project managers can raise issues with the program CCB to ensure that sufficient time, budget, and resources are available and that the change will increase the overall benefits for the program. Remember that although a project is assigned a budget, the ownership of the program budget and schedule falls on the shoulders of the program manager and must be managed across multiple projects and operational tasks.

Unexpected cost overruns – This is a very common problem regardless of the size of the project or program. One technique that we use is EAV to regularly measure project performance as it relates to schedule and cost. Through this tool we are able to forecast out estimates to completion (ETCs) and estimates at completion (EACs). When a project is running over budget, there are a number of ways to help the project. First, we triage the problems that are being encountered. Are they technical, training, or performance issues? Has scope creep actually occurred? Are there environmental factors that are creating delays in the project? From this triage we can develop strategies to overcome issues.

Poor communications – Communications is critical to both program and project management. To be successful teams must be comfortable discussing risks, concerns, technical issues, and potential technical issues that are being faced. In addition, project managers must be able to communicate with program managers honestly and openly regarding project status, risks, and resource needs. One of the ways to overcome these issues is to create a positive environment for open communication, avoid negative feedback and critical comments, and hold regular status meetings to discuss project issues. All team members should feel comfortable discussing concerns and issues including quality, technical issues, and unknown risks that have been realized. As a program manager, the goal is to ensure that each project manager creates a positive environment within his or her teams and that the program manager is involved and open to communication from any team member regardless of which project he or she is working on.

I worked on a program where team members held back quality concerns until the last minute before deployment. When these issues were revealed, we had to delay the release of the project until the quality issues were resolved. To promote earlier communications and more open discussions of concerns, we rewarded the individuals who identified the quality concerns in a public and open manner, creating an environment where team members were more comfortable exposing concerns early and often.

Bad or ineffective leaders – Hopefully, this book will assist you in becoming a better leader as a program manager, but quite often project managers can create their own negative environments and can be ineffective in the management of their teams. To avoid issues such as this, program managers should be included in project team meetings where possible, have stand-up meetings with team members, and where negative issues exist, address the project manager directly with a positive conversation about how his or her management style is negatively affecting the team. It is not uncommon to find project managers who either create negative environments or do not build up the confidence of the project team members.

I had one project manager who was well versed in project management processes but consistently held meetings without agendas, did not build confidence within the team, and had less than effective understanding of the project and the technology being leveraged. To assist the project manager, I started attending project status meetings and identified concerns with the way meetings were held and how the confidence of the team was waning. I then addressed the project manager regarding concerns and provided advice as to how to be more effective in running the meetings. I also assigned a senior developer to work with the project manager to better explain the software and the technology that was being used. The developer assigned helped to explain to the project manager the impact of comments and risks that were being identified in the meetings, translating the technical issues into an understandable level for the project manager. Very soon, the project manager was able to start to build the confidence of the team members by demonstrating a better understanding of the concerns and leading meetings with a greater confidence level. With the understanding he gained, he was much more confident in the project, the team, and the goals of the project.

Technology changes – Changes in technology midstream can destroy project schedules, cause team members to struggle with learning curves, and create new technical issues. Furthermore, the new technology can create risks that were previously unidentified and will often require the project manager to have to start over with project management plans, work breakdown structures, and risk management. The most effective way to handle changes such as this is to leverage EVA to demonstrate the investment already made in the project and ETCs to show the costs of completing the project maintaining the current technology versus the costs of implementing new technology. Most executives and program sponsors will understand the ROI and the costs of change, causing them to rethink the requirement to modify the technology approach.

However, there are times when the new technology is required because of integration with other products, improvements in approaches, or because a product will not be supported going forward. In cases like this, the CCB must be consulted to approve the new schedule and costs, and the project plan must be re-baselined to meet the requirements of the new technology.

From a program management level, the changes to a project's technology will directly affect benefit realization plans as well as program deliverables and deployment timings between multiple projects. The adaptation of a new technology for one project may require all projects to modify their technology, further expanding the cost and schedule issues. In cases like this, the program CCB must be consulted to discuss and approve the technology changes and the potential impact to the program. Oftentimes, the CCB can overcome the technology changes based on benefit realization, project schedules, and overall program budget.

Key management changes – Management changes should be avoided at all costs. Project managers build rapport with their teams, and switching project managers midstream can cause tension and anxiety within the team. New project managers will have to come up to speed on the project, establish new relationships, and may leverage different leadership styles. If problems are occurring with a project manager, the program manager is responsible for helping the project manager to be more effective. As mentioned earlier, helping the manager in leading teams, gaining technology knowledge, and building rapport with the team can be done with program management support. However, if this is not sufficient, assign a project coordinator capable of achieving better coordination between team members as well as being responsible for scheduling, setting meeting agendas, taking meeting minutes, and sorting through issues and risks to ensure that the project manager can better focus on team development, the deliverable schedules, and cost analysis.

In the event that a project manager must be replaced, the program manager can help to smooth the process by taking the team through the five stages of team development, providing continuity for the project, and ensuring that open communication is maintained.

Attrition – Although we try to avoid attrition of team members, it is a normal part of every team. Most often the best team members are the ones who find new opportunities and new challenges. Open communication with team members will enable management to be aware of team member concerns such as salary, new challenges, concerns with the organization, and relationship issues. Through open communication, program managers and project managers can attempt to address these issues before team members begin their search for new opportunities.

Quite often I have had team members who were tired of the legacy technology they were working with and began looking for opportunities where they could learn new technology and remain competitive in the marketplace. To overcome these concerns, I attempt to gain an understanding of the technology that the individual wants to work with, potentially provide training or offer seminars in the technology, or assign the team member to spend time researching the technology to see how it can be used in projects moving forward and they could leverage the technology of gaining the skills they sought.

Not managing stakeholder expectations – Stakeholder management falls on both the program manager and the project manager. Each project will have a set of stakeholders who will be influenced by the project and must be part of the project to ensure that they understand the project approaches, advantages, and delivery schedules. At the same time, the program manager is responsible for managing stakeholder communication across all programs to ensure that benefits are clearly identified and as benefits are realized from the completion of a project.

In addition, establishing success factors at the beginning of the program will go a long way to managing stakeholder expectations. By discussing and documenting the benefits that the program will achieve along with the timing of deliverables, stakeholders become part of the conversation establishing the scope and what to expect from program benefits and deliverables.

The next few sections outline HPT success factors that can be quite useful in building HPTs and ensuring that the teams stay together as a truly effective team meeting objectives, schedules, costs, and quality.

HPT SUCCESS FACTORS

Shared Purpose and Direction

> Leadership is intentional influence.
>
> —**Michael McKinney**

On an HPT, everyone on the team is committed to not only the project goals but also the program objectives, benefits, and strategic objectives. They know exactly what the vision of the program is because the program manager and project managers keep them focused by constantly communicating that purpose and vision in team meetings and providing regular updates on project/program progress, corporate strategic objectives, the achievement of benefits, and the program risk exposure. The leader helps each individual team member understand his or her contribution to the strategic objective to better enable team members to meet their own needs while serving the overall purpose of the team.

To achieve this, the program manager must have a clear agenda for all meetings and communicate early and often the program vision, objectives, and benefits. Program managers should be involved in project meetings on a regular basis and interact with project team members to ensure that they are clear about all levels of the program in their understanding of program/project success. While the program manager cannot undermine the leadership of the project manager, they will need to interact with each other, cross-check that (1) team members are receiving a consistent message, (2) risks identified at the program level are being communicated downward, and (3) project risks are being communicated up to the program level.

Motivating Goals

> "You cannot be a leader, and ask other people to follow you, unless you know how to follow, too."
>
> —**Sam Rayburn**

The program manager ensures that everyone on the team has clearly defined goals and targets recognizing the value of their contribution. In some organizations, the strategic goals and departmental objectives are determined by senior management. The program manager makes sure that these goals are clearly communicated and that each team member understands the value to the organization when program benefits are achieved. Team members should understand how their jobs support the achievement of the defined goals and, if possible, have the opportunity to develop individual goals and action plans that spell out how they will contribute to the success of the organization. Goals should also be measured and reported on to ensure that team members can recognize their achievements and that leadership teams can reward success through acknowledgment and public recognition.

Commitment to Individual and Team Roles

> "The function of a leader within any institution: to provide that regulation through his or her non-anxious, self-defined presence."
>
> **—Edwin H. Friedman, *A Failure of Nerve***

On an HPT, members have clearly defined expectations, but they also understand how each of their roles is linked to every other role. Program managers and project managers ensure that team members are cross-trained in other responsibilities and understand the total effort so that everyone is aware of the overall objective, enabling them to back each other up when needed. The program and project managers make sure that individual job responsibilities are fulfilled, but at the same time work to help all members develop a common language, processes, and approaches that allow them to function as a team. Industry best practices such as PMI approaches often lay the foundation for methodology approaches, and each team should have a RAM or a RACI chart to ensure that all team members understand their contributions and their roles in contributing to the success of the effort.

Multidirectional Communication

> If I had to name a single all-purpose instrument of leadership, it would be communication.
>
> **—John Gardner**

On the best teams, team members work closely together to solve problems, communicate openly with each other, and keep the program and project managers updated on current challenges or emerging issues. On low-performing teams, communication is one-way (from management and the organization to team members). Skilled leaders focus on developing multidirectional communication, avoiding the trap of only communicating out to individual members of the team and not listening. Communication only with some individuals can often leave other team members out of the loop and can cause frustration and confusion over project objectives.

To ensure open communication, leaders will often raise discussion points and allow team members to continue the conversation with very little interaction from the leader. Opening and encouraging these communication channels helps teams develop better relationships and communication skills that can be leveraged outside of a formal meeting. In addition, vocally supporting all positions helps team members to see the value of their contribution and get public recognition for their opinions. Facilitating meetings in this way transfers much of the power from the leader to the team members and opens better communication channels. At times, conflict will arise requiring the leader to step in if the team turns negative, vocalizing rules for conflict discussion and supporting opposing viewpoints. Conflict resolution (Chapter 15) is one of the key success factors that make a team move from one that meets expectations to an HPT that is capable of exceeding expectations.

Good communication, starting with the program manager, is one of the key factors in establishing team success. Communication increases commitment from all members and establishes connections/relationships with team members and leaders. For HPTs all team members must be able to talk with and listen to each other without issues.

Teams face dilemmas when they are not communicating with each other and with management. If individual team members are not aware of what is being accomplished, productivity can come to a halt because no one understands the agenda and accomplishments that have been achieved. In the 1991 book *Empowered Teams*, Richard Wellins, William Byham, and Jeanne Wilson state that "communication refers to the style and extent of interactions both among members and between members and those outside the teams. It also refers to the way that members handle conflict, decision making, and day-to-day interactions" (3).

Leaders leverage communication with teams, stakeholders, management, and organizations to ensure that the efforts they are leading will achieve success and that expectations are set for the benefits to be delivered. As a program manager, the three most critical aspects for communication are consistency, clarity, and courtesy:

Consistency – Leaders who vary in the messages sent to teams confuse and frustrate them, leading them to have less respect and often concern that the leader is not communicating the truth and misleading the team. Program leaders must remain consistent in communication, ensuring that messages to one project manager or team are consistent with messages sent to another. In addition, the program manager must maintain that same message throughout, and if the message needs to change, he or she must communicate the change and the reasoning behind it. Without this, teams cannot achieve the level of productivity necessary for HPTs.

Clarity – Teams and individuals cannot execute on messages that are not clear and consistent. Program managers need to ensure that the messages being transmitted are not only stated consistently but also understood completely for team members to be able to achieve desired results. Straightforward and clear messages ensure that all listeners can understand the messages and achieve the desired results or actions.

Courteous – Regardless of whether he or she is dealing with a team member or an executive in the organization, an effective program manager will demonstrate a level of respect. Everyone deserves to be treated with respect, and members of an HPT will perform much more effectively among themselves if the leader sets the tone of mutual respect.

"Few people are successful unless a lot of other people want them to be."

—Charlie Brower

Teams must also be able to communicate openly and honestly among themselves. In an HPT, all team members need to communicate for the common good of the program. Therefore HPTs demonstrate supportive, active, and vulnerable levels of communication. They actively work to communicate, support each other's ability to state what is important, and not posture with one another. While no idea is a "dumb idea," many teams restrict communication and self-monitor their own messages until they are well thought out and supported by a basis of fact. In an HPT, team members are able to discuss issues without barriers to communication and need to be free to suggest "dumb ideas" even if the idea really didn't add value to the effort.

Finally, team members need to understand that the program manager wants, and needs thorough open and honest communication. No effective program manager will work with "yes men" who agree to everything the leader says. They are not contributing to the discussion and work against a team's ability to trust one another. An old saying is that "if two people agree on everything, one is not needed"; in other words, controversy and conversation require discourse and differing opinions. Program managers need to be very careful when working with teams to ensure that they encourage disagreement and open discussion to better gain insight and value of external opinions. Otherwise some of the very best ideas can be missed and team members will not contribute effectively to solving problems.

Authority to Decide or Act

> Leadership is the ability to establish standards and manage a creative climate where people are self-motivated toward the mastery of long term constructive goals, in a participatory environment of mutual respect, compatible with personal values.
>
> **—Mike Vance**

New teams have to earn this autonomous authority by demonstrating that they understand the team's purpose, processes, and priorities. However, effective team leaders work toward providing opportunities for team members to earn this trust, pushing authority for team decision making to the team members. Team members know how and when to get approval for decisions and, in the best of cases, are charged with making on-the-spot decisions when faced with issues. Only critical issues with conflicts are escalated to the leadership team for resolution. On low-performing teams, team members have to constantly get approval before taking action, significantly reducing their effectiveness and negatively affecting their sense of engagement on the team.

However, regardless of the process employed, as a general rule an HPT will operate with a deeper sense of shared purpose, attempt to achieve more ambitious performance goals than their counterparts, have better communication across the team, and demonstrate a shared set of respect and accountability toward their teammates. This is not accomplished easily.

If you have led a team, you know that achieving high performance is ongoing. Rarely does a team that achieves high performance just stay at that level without leadership and intervention. Quite often teams are short lived, focusing on the objectives of the project or program and then disbanded for other efforts. Each program or project will have a different set of team members with their own experience and skill sets; therefore every program and project must go through the five stages of team development and be encouraged to establish trust, communication, and positive conflict resolution.

Reliance on Diverse Talents

All leadership is influence.

— John C. Maxwell, INJOY

Effective team leaders pay attention to helping team members understand their unique strengths, talents, and weaknesses. No individual team member can be good at everything. The best leaders help everyone to develop an appreciation for individual style differences, natural gifts, and personal experience. Teams are encouraged to use the language of acceptance and appreciation rather than criticism and judgment. Leaders consciously hire team members who bring complementary skill sets, unique experiences, and diverse perspectives. Teams are taught to use positive conflict resolution instead of negative and derogatory interaction, creating a safe place for open and honest communication and encouraging each other to innovate, identify risks, and be as creative as possible in driving the effort forward.

A program manager's responsibility is to ensure that teams are offered the opportunity to work together to achieve this level of communication and trust, while empowering project managers to drive teams to their highest-performing standards. Program managers must often clear hurdles for teams to be able to build into an HPT while also helping the project manager to focus on team building, opening communication channels, and creating a comfortable area for the exchange of ideas, concerns, and risks.

Mutual Support and Trust

> The final test of a leader is that he leaves behind him in other men the conviction and the will to carry on.
>
> **—Walter Lippmann**

The team leader can't force a team to be supportive and trusting—it's a natural result of positive conflict resolution, open communication, positive environmental factors, shared responsibility, and mutual respect. The HPT achieves mutual support and trust because they have a history of working together to achieve grand dreams and results. They have met challenges, overcome obstacles, and backed each other up in good times and bad. The leadership and team have earned each other's trust in a positive working environment supported by management and empowered to succeed.

Yet there are a number of tools that a leader can bring to bear in facilitating HPTs. Opening communication channels, working with team members to resolve conflicts from previous interactions, forming trust between management and teams, and having team members work on tasks together to gain respect for each other's skills can all be beneficial. Working with the team is always a unique experience, and depending on the individuals, employing differing leadership styles will help to contribute to overall success. Each member of a team has a different view of the organization, culture, and management. Therefore each program manager must work to understand the individuals and develop concrete strategies to bring them toward a common understanding and goals for contributing to the success of the team.

Building an HPT is not an easy task. However, if you're a program manager that is up to the challenge, then consciously focus on developing HPTs is a primary responsibility, resulting in a more positive working environment and increasing the potential for successful program delivery.

CHARACTERISTICS OF A HIGH-PERFORMING TEAM

> Your position never gives you the right to command. It only imposes on you the duty of so living your life that others may receive your orders without being humiliated.
>
> **—Dag Hammarskjöld**

Almost all HPTs have certain key characteristics in common. Continuously analyzing and evaluating the team based on these criteria is critical in order to keep the team dynamic and effective. Here are the top key characteristics of an HPT:

A team requires a clearly stated purpose and goals—not just an understanding of what needs to be done at the moment, but an understanding of the overall focus of the team. Shared goals and objectives lead to commitments. Members of HPTs share a sense of common purpose. They are clear about the team's "work" and why it is important. They can describe what the team intends to achieve and have developed mutually agreed upon and challenging goals that clearly relate to the team's vision. Strategies for achieving goals are clear. Each member understands his or her role in realizing the vision.

Experimentation and creativity are vital signs of a dynamic team. Dynamic teams take calculated risks by trying different ways of doing things. They aren't afraid of failure, and they look for opportunities to implement new processes and techniques. They are also flexible and creative when dealing with problems and making decisions.

The ability to produce what is required when it is required is a true test. A dynamic team is capable of achieving results beyond the sum of its individuals. There is a commitment to high standards and quality results. They get the job done, meet deadlines, and achieve goals. Members have developed strong skills in group processes as well as task accomplishment. Team members continually meet time, budget, and quality commitments.

A dynamic team clarifies roles and responsibilities for all its members. Each team member knows what is expected and knows the roles of fellow team members. A dynamic team updates its roles and responsibilities to keep up with changing demands, objectives, and technologies. Team leaders ensure that each member is cross-trained in other responsibilities so that everyone can support each other when needed. The team leader makes sure that individual job responsibilities are fulfilled, but at the same time works to help individuals develop a common language, processes, and approaches that allow them to function as a team.

An HPT defines protocols, procedures, and policies from the very beginning. Structure allows a team to meet the demands of any tasks it must handle. Information is easy to access and available at all times for the team members.

Leaders of HPTs regularly catalog their team's knowledge, skills, and talents. Team leaders are aware of their members' strengths and weaknesses, so they can effectively draw upon individual competencies. There is an appreciation for individual style differences, natural gifts, and personal experience. The members of the team are encouraged to use the language of acceptance and appreciation rather than criticism and judgment. The team leader consciously hires team members who bring complementary skill sets, unique experiences, and diverse perspectives.

Dynamic teams share leadership roles among members. Such teams give every member the opportunity to "shine" as the leader. Team members also appreciate formal supervisory roles, because the formal leaders of a dynamic team support team efforts and respect individual uniqueness.

An HPT has members who enthusiastically work well together with a high degree of involvement and group energy (synergy). Collectively, individual members feel more productive and find that team activities renew their interest and spirit. Such a team develops a distinct character of its own.

Disagreements will occur in all teams. It's not necessarily negative or destructive. A dynamic team deals openly with conflict and tries to resolve it through honest discussion tempered by mutual trust.

Members of a dynamic team talk to each other directly and honestly. The team is committed to open communication, and members feel they can state their opinions, thoughts, and feelings without fear. Listening is considered as important as speaking. Differences of opinion and perspective are valued and methods of managing conflict are understood. Each person solicits suggestions from other members, fully considers what is said, and then builds on those ideas. Through honest and caring feedback, members are aware of their strengths and weaknesses as team members. There is an atmosphere of trust and acceptance and a sense of community in a team. Group cohesion is very high.

Dynamic teams have well-established, proactive approaches to solving problems and making decisions. Decisions are reached through consensus; everybody must be able to "live with" and willingly support decisions. Members feel free to express their feelings about any decision. Team members clearly understand and accept all decisions, and they develop contingency plans.

A team needs to routinely examine itself to see how it's doing. "Continuous improvement" and "proactive management" are operating philosophies of dynamic teams. If performance problems arise, they can be resolved before they become serious.

An HPT has effective, productive, well-managed meetings that efficiently use team members' time. Every meeting is focused, timely, and necessary, and is used to solve problems, make decisions, disseminate information, and enhance team member skills.

In an HPT, individual and team accomplishments are frequently recognized by the team leader as well as by team members. The team celebrates milestones, accomplishments, and events. Team accomplishments are also noticed and valued by the larger organization.

Members in HPTs are enthusiastic about the work of the team, and each person feels pride in being a member of the team. They are confident, committed, and optimistic about the future. There is a sense of excitement about individual and team accomplishments and the way team members work together. Team spirit is high.

HPTs know how use conflict to build the team instead of destroying it:

- An HPT develops goals and a mission statement from the beginning of the project. Each member of the team has a specific job title and purpose in accomplishing those goals. The goals help to provide direction and ensure productivity in a timely manner.
- Open communication is a common characteristic of an HPT. Open communication can include feedback or brainstorming sessions, surveys, and discussion and focus groups. The individuals in an HPT are encouraged to share their thoughts, feelings, and suggestions with one another.
- Conflict is a part of team dynamics. Successful HPTs know how use conflict to build the team instead of destroying it. Each member is committed to developing and maintaining positive relationships through nonaggressive confrontation and verbal and nonverbal communication. The team members show one another respect and work together for the common good of the team.
- Problem solving is a large aspect of HPTs. The team learns the strengths and weaknesses of each team member and is able to capitalize on different team members' abilities when needed. Problem solving begins with a team effectively identifying where the problem began and how to repair it as efficiently as possible.
- Effective HPTs are able to respond to and respect their leader. For example, when an executive decision needs to be made and the leader decides to implement a certain strategy, the team members respond immediately with support and action.

- HPTs are given the opportunity of training and developing in specific areas such as leadership principles, organizational effectiveness, and communication skills. Training and development manifests itself in formal classes, training sessions, and resources such as books and personal coaching.

SETTING GOALS THAT INSPIRE HIGH TEAM PERFORMANCE

> Humans will probably always need the help of especially gifted moral leaders in order to extend the bonds of caring and trust beyond the easy range of the family and the face-to-face community. Such bonds have become essential to the future of humanity.
>
> **—Paul R. Lawrence**
> *Driven to Lead*

Teams function based on how they are tasked and what goals or objectives are set for them. To achieve an HPT, goals must be set that challenge the team to achieve but also have a clear scope and guidance so the team can focus their efforts in the direction best suited for delivery. Objectives for a team need to be clear, concise, and measureable so both the team and management can gauge successful progress and achievement. It is a fine balancing act to set a goal that is broad enough that it requires innovation and challenges the team but narrow enough in scope to ensure that the team can make decisions and define approaches to solving the issues. If a team is given a clear-cut goal such as mow that lawn, there is no room for creativity and innovation. The progress can be measured and the effort may require multiple people to achieve it, but the team is not achieving its potential because the direction has been laid out for them. If on the other hand the team were given a goal such as maintain the greenery in a cost-effective manner that lasts long term but still meets acceptable criteria, the options for the team expand and the creativity of the team can be leveraged. Goal-setting theory (Locke and Latham 2005) provides a foundation for leaders to develop goals that challenge individuals and teams.

Goal-Setting Theory

Never give an order that can't be obeyed.

—General Douglas MacArthur

Dr. Edwin Locke's (1968) research paper "Toward a Theory of Task Motivation and Incentives" outlines the foundation we use today to set goals for teams. Locke determined that there was a direct relationship between the performance of individuals and the complexity of the goal set. Goals that were too easily achieved were not a challenge for people and therefore their personal performance decreased. The more complex a goal, the harder people worked to achieve the goal. In addition, the more concise a goal, the more likely it was that a team would achieve it, whereas a vague goal that is ambiguous and difficult to understand makes it less likely that the team will achieve it. A specific but challenging goal will drive greater performance than a vague or easily achieved one will.

Teams working toward a difficult challenge will often rise to the occasion and deliver higher performance. Engaging teams to achieve goals is a major challenge for program managers. If the answer were simple and easily achievable, the odds are that a program would not be undertaken to find the answer. Therefore the program manager, looking across all component projects and tasks, needs to develop goals that challenge teams but are understood and have clear, concise objectives. Goals have to be specific, not general:

- Goals have to be challenging, yet realistic.
- Goals should be established through the SMART methodology:
    - Specific
    - Measureable
    - Achievable
    - Realistic
    - Timely

Locke's theory states that simply asking for high performance is not sufficient to achieve success or HPTs. Instead, a challenge must be presented that engages the team to overcome it and to strive to achieve the objectives. High team performance is a lot more probable when the goals are specific and present a challenge that is perceived to be realistic and attainable by the team.

In many cases, HPTs are the ones that undertake the most challenging of efforts and through the challenge the team begins to work as a cohesive whole, relying on each other's skill sets to achieve objectives and overcome the challenges. Very rarely do HPTs exist when the tasks are simple, ambiguous, or simply not achievable. Team members will put in the requisite efforts to "go through the motions" on a program where the objectives are not feasible, but the same team members will give their all for an initiative that challenges them but, while a stretch, is achievable. HPTs operate best when challenged and can see the potential for success.

A program manager leverages program vision, project deliverables, and benefit realization plans to present challenges to teams that they can understand and drive forward to achieve success. The effective program manager will work with project teams to lay out the challenges, risks, and long-term objectives, encouraging the teams to assist in developing strategies to meet the challenge and engaging teams to determine approaches to overcome risks.

Performance Orientation

> My own definition of leadership is this: The capacity and the will to rally men and women to a common purpose and the character which inspires confidence.
>
> **—General Bernard Montgomery**

Teams respond with greater performance when goals are challenging and specific, but are truly inspired when the challenge is one that will achieve tangible benefits but would not be easily produced by going through the motions. An inspiring goal is one that challenges the team and piques their own competitive nature, invoking personal pride and motivation. Yet at the same time, without a way by which the team can be measured, their effectiveness will be unrecognized. It is tough to root for a sports team without scores being kept, and it is difficult to support races with no winner. Teams need to have not only a challenge set in front of them but also a manner by which their accomplishments can be measured, demonstrating their progress and achievements. Any team setting out on a task will do better when the results and performance expected of them are clear and publicly recognized. Performance orientation is the use of goal setting to link team goals to organization performance. When the goal is set to align with organizational objectives, and the teams can measure their own contribution, the team has the ability to become a truly HPT that strives to achieve a level of success beyond what a nonperforming team can achieve. Invoking the competitive spirit of individuals motivates them to push further toward a goal. Setting team objectives enables the team to form together and rely on one another to achieve success. And defining performance measures that can be used to demonstrate their achievement not only enables teams to meet objectives, but often drives them further in personal and group performance.

14

Improve Team Motivation, Morale, and Productivity

> When the conduct of men is designed to be influenced, persuasion, kind, unassuming persuasion, should ever be adopted. It is an old and a true maxim, that a "drop of honey catches more flies than a gallon of gall."
>
> **—Abraham Lincoln**

It still amazes me how few organizations invest in understanding and improving the morale of their staff. Companies that spend an inordinate number of dollars advertising, providing customer service, warranty fulfillment, and doing anything to retain customers can be completely obtuse about the morale of the teams that work for them. Only when situations such as extreme attrition or complaints about management are escalated do they seem to recognize that they have an entire staff that needs to be kept motivated, productive, and ultimately happy.

Low morale in an organization leads to poor cooperation, low productivity, and increased attrition rates. Organizations suffering from low morale will often be places where gossip runs rampant (especially anything negative), productivity is poor, efficiency is low or nonexistent, and staff are resentful of management. One sure way to identify an organization suffering from low morale is to monitor the absentee rates. Those with very high absentee rates are most commonly subject to equivalently high attrition rates, requiring the organization to spend time and money hiring and training new staff members.

Regardless of the organizational culture, a program manager needs a team that is motivated, challenged, and positive about the work that they are achieving. The program manager needs to create a bubble in which the team is protected from the negative cultural issues and encouraged to participate in the positives of the culture. They must understand the value of the program and want the organization to succeed through their contributions. That's not an easy task for a program manager, especially when the team exists in a hostile or toxic environment, but a requirement if the program is to achieve its objectives in a timely and cost-effective manner.

Some of the tools that I have leveraged include:

- Communicate with teams that success is more than just their job.
- Provide positive recognition for accomplishments.
- Encourage staff to work on tasks that they are passionate about.
- Set a series of internal processes regardless of how the organization traditionally operates.
- Set an example by having fun with your daily routine—employ humor and a smile whenever possible.
- Encourage staff to be positive.
- Reward extra work and effort with compensation or time away from the office.
- Protect the team from extraneous negativity.
- Communicate openly and honestly at all times.

Sometimes it is the smallest efforts that have the biggest gains. Rewarding people, recognizing their successes, and demonstrating a positive attitude regardless of the situation all demonstrate a set of traits that are infectious to the team and quite often to the organization around you. A positive and healthy team is one that other staff members will want to join and one that team members can take a certain pride in. When the organization attempts to invade the positive environment you have created, it is the program manager's responsibility to interject and protect the team from any negativity.

One final point, though some of my peers may disagree with me on this, is that communicating honestly and openly with the team is a critical success factor. For some reason managers quite often keep vital information away from team members, preventing them from rising to new challenges or being able to create innovative strategies. Regardless of the positive environment that you as a program manager have created, the rumor mill will be running and any bad news will circulate through the organization. If you have invested time and energy into building an HPT and creating a positive environment, don't destroy all of the hard work achieved by misrepresenting or lying to the team. Ensure that open communication is used for both positive and negative news, and leverage the team to contribute to success through the skills and abilities that they have.

As an example, I worked with a team that developed software, leveraging a particular technology-based language. I had attended the annual conference for the technology and listened to the CEO announce that they would discontinue the language moving forward. Completely demoralized, I looked at all of the legacy software that had been built on the language and the new efforts that were on the table for development. The decision by the firm to discontinue the language had a huge impact on employees as they had been trained and perfected their skills in what was now an obsolete approach.

I took the news back to the team and opened the conversation up for discussion on strategies that we could employ. One team member suggested that we attend a conference the following month where Microsoft was announcing their new technology. I took the entire team to the conference and met daily with them to discuss the new approach and determine how we could leverage what had been done against the approaches required for the new technology. The team came together and developed strategies for team members to educate themselves, learn the new approach, and develop approaches to migrate the old code into the new technology.

Although the overall program was a lengthy one requiring a tremendous amount of learning, team interaction, and in some cases trial and error, the team quickly came up to speed and converted legacy applications into the new technology while also taking on new development efforts. After three years, the team had achieved a 0 percent attrition rate, worked so closely together that it was difficult to separate individuals, and had one of the highest morale levels I had ever seen.

The goal was clear and concise, performance was measured by both the conversion from legacy into new technology as well as the creation of new products, and the team outperformed any of its counterparts, setting the standard for the organization as a whole. A true HPT was achieved by open communication, taking some of the worst news I have encountered in the software industry, and turning a negative into a positive result with tangible achievements.

15

Conflict Resolution

A leader is a dealer in hope.

—Napoleon Bonaparte

Conflict is a normal part of everyday life, whether work or personal. In our personal lives we encounter conflict with friends, strangers, aggressive drivers, spouses, and children. Each conflict we face can escalate to a negative level, affecting long-term relationships or short-term interactions. With conflict playing such a regular role in everyday life, it would seem that we would naturally develop techniques to avoid or minimize conflict and to avoid it where possible. Yet conflict is a natural part of our psyche whether it is differences in politics, national conflict, sports teams, or even which TV show a family will turn on (see Figure 15.1).

Therefore it is not surprising that we encounter conflict regularly in work-related activities. Interacting with management, coworkers, clients, and vendors can quickly degrade into negative relationships between people and teams. Negative conflict involves degrading the other party, minimizing their opinions, avoiding conversations, and at times exchanging harsh words that can have a dramatically negative, long-term effect on relationships.

Yet one of the many myths about conflict is that it is only negative. Conflict can be very positive when managed in a way where respect for one another allows for the open discourse of ideas, perceptions, and opinions. Positive conflict results in greater innovation, greater creativity, and the combination of ideas toward the betterment of the goals and objectives. Positive conflict builds relationships and mutual respect between team members and facilitates a greater exchange of ideas beyond simply technical concerns.

The Nature of Conflicts

Conflict is natural, neither positive nor negative, it just is.

Conflict is just an interference pattern of energy.

Nature uses conflict as its primary motivator for change, creating beautiful beaches, canyons, mountains and pearls.

It's not whether you have conflict in your life, it's what you do with that conflict that makes a difference.

Conflict is not a contest.

Winning and losing are goals for games, not for conflicts.

Learning, growing, and cooperating are goals for resolving conflicts.

Resolving conflict is rarely about who is right, it is about acknowledgment and appreciation of differences (Crum and Denver 1987, p. 49).

FIGURE 15.1
Nature of conflicts.

As a program manager, building an HPT requires creating positive conflict resolution skills among team members and creating a safe and secure environment for the exchange of ideas, concerns, and questions. A program manager ensures that conflict does not break down into negative interaction, derogatory comments, or the shutting down of individuals. Rather, the program manager must make sure that every opinion is heard, discussed, and evaluated for potential risks, advantages, and opportunities (see Figure 15.2). Although not every opinion offered by a team member will result in a change to the program or become a resolution to a challenge, team members must feel that they have a positive and safe environment for the exchange of ideas.

Through a positive conflict resolution process, relationships between team members are made stronger, communication channels open up further, and trust is built between team members. Further, the leadership team is supported in empowering staff to discuss new ideas, concepts, strategies, risks, and concerns. However, achieving positive conflict resolution is a challenging objective and one that requires constant monitoring and intervention.

Although the program manager is generally responsible for working with the program office personnel and project managers, there are times when they also must work with individual project teams to help or facilitate conflict resolution processes. All teams, regardless of their past experience, will have to establish processes and communication channels that facilitate proper conflict resolution. As such, it falls on the program manager to ensure that proper conflict resolution techniques are in place, followed,

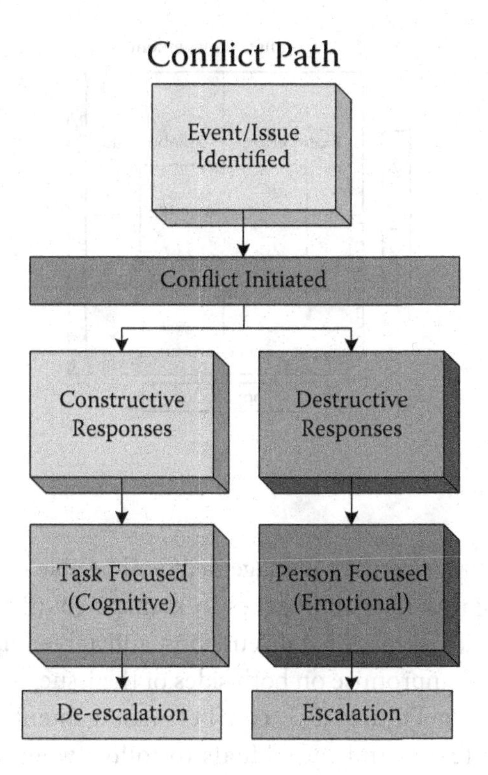

FIGURE 15.2
Conflict path.

and employed regardless of the uniqueness of the situation or personalities involved. A team that can resolve conflict works in a positive and safe environment where each individual contributes to the overall success of the effort, whereas teams that have negative conflict result in personality differences, negativity, escalation of issues, and hostility. Obviously, an HPT will need to have positive conflict resolution to be able to work together effectively without management intervention when issues arise. Unresolved and ineffective conflict management will follow five stages:

1. Discomfort
2. Incident
3. Misunderstanding
4. Tension
5. Crisis

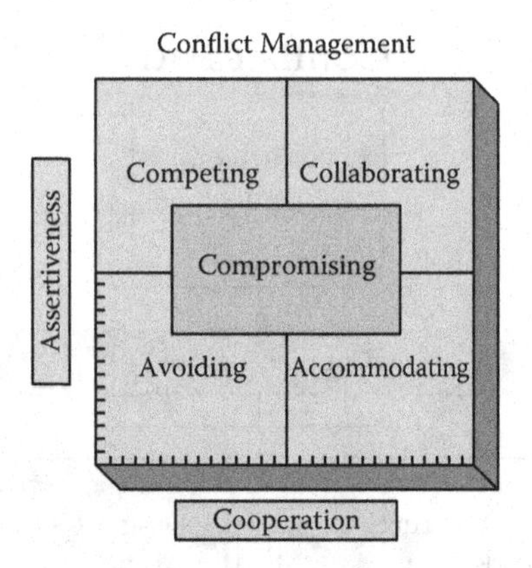

FIGURE 15.3
Conflict management.

A team with positive conflict management will be able to avoid a crisis by working through various approaches to manage conflict with constructive conversation, task-oriented discussions, and agreed upon resolutions, often forming a compromise on both sides of the issue.

As illustrated in Figure 15.3, conflict management focuses on five approaches for teams and individuals to follow when an issue arises. These five approaches range across a spectrum from cooperation to assertiveness, and when balanced, team members assert their beliefs while cooperating for a positive resolution. A balanced approach will result in both sides of a conflict understanding the perspective of the other side and will achieve an agreed upon solution without forcing escalation or management intervention. Balancing conflict resolution between assertiveness and cooperation enables all team members to participate in discussions, creates a collaborative environment, avoids negative attacks or personality conflicts, and ultimately results in the optimum solution because each side works together to deliver an effective solution. The most effective approach for achieving positive conflict resolution is one in which each individual is encouraged to participate in discussions, and constructive dialogues take place in a collaborative environment.

Other approaches can achieve a solution without management interaction, but can also create an environment where team members grow to believe that their opinions don't matter or that their participation in achieving the objectives is less important. On the lowest end of the spectrum, conflict is simply avoided and a negative environment builds. At the highest end, team members collaborate together to understand both sides of the conflict and work through developing a strategy that employs the most effective approach for both sides. In this form of conflict resolution, neither side will walk away feeling that the conflict was not properly handled or that they did not contribute to the best possible solution. Instead, the team reaches a resolution without feeling compromised, and the environment for conflict resolution is a positive one.

FIVE APPROACHES TO CONFLICT MANAGEMENT

1. **Avoiding** – In this scenario, individuals avoid directly confronting opposing points of view and sometimes avoid even opening the discussion, willingly giving in to the more assertive position without effectively communicating their position. While this will avoid conflict becoming personal or negative, it will leave individuals feeling that their opinions are going unheard and that their value to the team is less than the more assertive individuals. Avoiding will result in a less than positive environment and a less effective solution. Teams operating in these environments are not able to resolve conflict in a healthy way and will perform less effectively than HPTs.
2. **Accommodating** – Though effective in avoiding conflict, accommodation often results in one team member agreeing with another without pushing his or her individual approach or opinion. The solution is defined and agreed upon, but the individual accommodating the solution is not being heard effectively and there is a risk that the solution will lack an important point because of the lack of assertiveness of a team member. Accommodating may result in an environment where there is less direct conflict, but the team dynamics will be less positive and team members will operate as if their opinions are not valued or contribute to the program success.
3. **Competing** – Individuals assert their belief and argue for their position without participating in an open and honest dialogue. Their focus is on their proposed solution or strategy, and there is not a demonstrated understanding or valuation placed on the opposing side. Oftentimes management intervention is required to overcome competing conflict resolutions because neither side will accept the other's position. Competing conflict resolution will leave teams in a

more directly hostile environment and result in greater management intervention, potentially leaving other team members choosing to avoid conflict rather than offering their ideas. Teams with competing conflict resolution strategies create a more negative environment and struggle to achieve an HPT status.

4. **Collaborating** – The most positive of all conflict resolution strategies, collaborating on solutions creates a positive environment where each individual is provided with an opportunity to offer his or her position and concerns and identify potential risks. Each member is assertive in his or her approach but also willing to cooperate to achieve a resolution that both meets the needs of the program and includes all approaches. Management intervention is not required as the team works together through issues and develops strategies to solve conflicts.
5. **Compromising** – Though compromising will achieve a solution and facilitates each side being heard, the solution is generally a lose/lose for both sides. Each side of the conflict gives up some piece to the other so that a compromise can be achieved, and each walks away from the conflict with a less than positive result (see Figure 15.4). There are times when management intervention is necessary in compromising strategies, but as a general rule the team works to resolve conflicts on their own. While compromising does not directly cause a hostile environment, it can leave team members feeling that their positions were not effectively heard and that a less than perfect outcome was reached. Over time, team members using compromising strategies are more likely to move toward accommodating if others on the team are more assertive.

The most effective HPTs will work in a collaborative environment and have a positive conflict resolution approach, ensuring that issues are resolved in the most effective manner for the program and that individual interests are not the driving force. As a general rule, there is always a point of mutual interest that conflict resolution strategies should drive toward. To achieve this, a program manager will work with the team to develop and maintain positive conflict resolution approaches and empower team members to resolve conflicts together without management intervention while continuing a focus on developing a positive and healthy culture for open discussion. Oftentimes this requires program managers to train

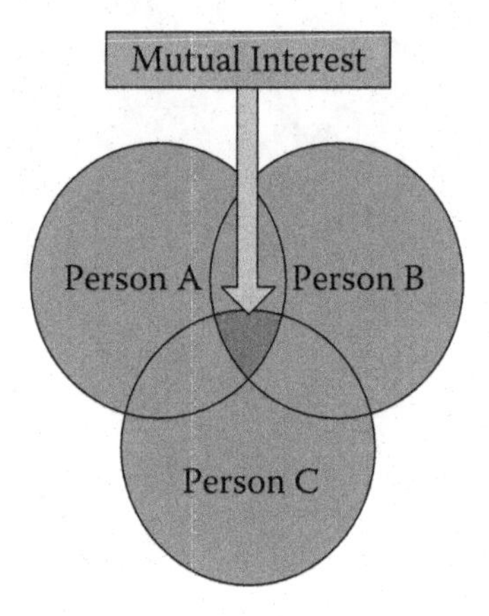

FIGURE 15.4
Common interests.

teams in approaches and occasionally intervene when individuals become too aggressive, focusing only on their own position to the detriment of the program objectives.

16

Case Study in Leading Teams

You have just been assigned as a program manager to a program that will require a five-year effort to increase the efficiency of a retirement service for a state agency. The effort will include cleansing data, buying or developing new software, implementing interfaces to multiple agencies for data transmission, and decreasing the time to process retirement payments from 120 days to 45 days. The program will involve process reengineering, technology teams, multiple state agencies, and very large database systems to manage data moving forward.

The first projects have been defined to cleanse the historic data and to create a data feed that will be leveraged to feed into the new system. Your task is to define the challenges and objectives for subsequent projects and to develop teams that will exceed performance and achieve the desired objectives. Each team will be made up of various skill sets and will have to contribute to the overall objectives of the program.

CASE STUDY QUESTIONS

1. What strategies would you use to create projects that challenge the teams with clear and concise goals?
2. How would you build HPTs with various groups when each group is taking on challenges ranging from business process reengineering to software development to building data centers?
3. How would you tailor your approaches to each project manager to ensure that a positive and healthy culture exists within their teams?

4. What level of involvement would you have for each project manager to contribute to the definition of new projects and their corresponding objectives?

DISCUSSION QUESTIONS

1. Building HPTs requires challenges, concise objectives, and a way by which the performance of the team can be measured. How would the approaches discussed be modified if the teams were spread throughout the United States? Would the approach differ if the team members were subcontractors?
2. Competition increases performance of teams. How do you as a program manager avoid competition within the team while still encouraging the team to compete against industry benchmarks or other departments in the organization?
3. If your team has issues in communicating, how would you as a program manager work to overcome the communication issues and move the team toward increasing performance and achieving a high level of performance?

Section IV

Formal Leadership Processes

17

Formal Processes

PROGRAM MANAGEMENT AS A FUNCTION OF PRODUCTIVITY

> High sentiments always win in the end. The leaders who offer blood, toil, tears and sweat always get more out of their followers than those who offer safety and a good time. When it comes to the pinch, human beings are heroic.
>
> **—George Orwell**

The focus of this book has been on program management leadership and the role that it plays in developing successful programs through better communication, leadership styles, team development, and organizational improvement. We have spent very little time covering the formalized process that PMI outlines, simply because there are many books on the process and very little on program management leadership. While there is plenty written on leadership as a whole, and on becoming that Fortune 500 CEO, program management leadership is a very different topic that deserves additional information added to the overall body of knowledge. While PMI has done a good job in outlining the process of program management, there are still areas that require additional knowledge and reference material to help extend the knowledge of program management. The role that a program manager plays in team productivity is a crucial one.

Program managers bring teams together, overcome obstacles for the effort, promote open and honest communications, and develop an environment of positive reinforcement for teams to achieve objectives. At the same time, they empower project managers to achieve the objectives of each project, work to overcome any obstacles, and often balance resources between projects, ensuring that specialized skills are available when needed as well as fully utilized across the program effort. Their responsibility is to see the "big picture" of the effort and facilitate project managers to focus on the more myopic outputs of the projects. It is the program manager who ensures that the projects are all aligned with the program benefits and will achieve the results desired.

A program manager leverages many of the general skills of management including general management, interpersonal skills, communication, financial management, human resource-related interactions, performance setting, and review, a program manager must also deliver leadership to be successful. In earlier chapters, we discussed building HPTs as part of a successful program effort. HPTs cannot be achieved doing mundane, task-only, or redundant work; instead they must be challenged, given objectives that are clear but difficult to achieve, form relationships and partnerships within the team, and operate with the full support of the program manager to enable innovation, creativity, and risk tolerance. HPTs are operating at their greatest level of performance when their competitive spirit is engaged and they are taking on challenges that are difficult but possible to achieve. The program manager should communicate the challenge, provide the tools and resources necessary to achieve it, communicate the vision, provide positive reinforcement, eliminate obstacles, and keep the team engaged throughout the life cycle of the program.

While all of these contributions are necessary for programs and teams to achieve success, program managers are primarily responsible for five main domains: (1) strategy alignment, (2) program benefits management, (3) program stakeholder engagement, (4) program governance, and (5) program life cycle management. Initially, a program manager will align the program effort with the strategic objectives of the organization. However, these strategy-based goals will change over time, and the program must constantly be evaluated to ensure that it maintains alignment with the direction of the organization. The program manager consistently evaluates program progress as it is aligned with the projected benefits to be realized by the program. As benefits are realized, the program manager ensures that the realization is communicated to the stakeholder community and that they clearly understand how the benefit realized impacts them as well as the longer-term intentions of the program.

The program manager also provides governance over all of the projects in the program to ensure that procedures and policies are in place and being followed. Governance is the creation and enforcement of process and policies to be followed consistently across projects and reported to the program manager, program team, and often a program governance body in a timely and effective manner. It provides an appropriate organizational structure with policies and procedures for project managers to operate under and a means to communicate their achievements, risks, schedules, and costs. Within this framework, the program managers can develop communication methods to provide status on complex projects and integrated deliverables as they work toward achieving overall benefits.

While governance provides the framework for communicating with projects, and stakeholder management ensures that stakeholders are well informed, the program manager drives productivity through benefit identification and ensuring that projects realize the defined benefits. If for some reason a project is not able to meet the benefits in the established time frame or at the cost established, the program manager has the resources available across all project lines to assist in adding resources, developing strategic plans, and leveraging the knowledge base of all projects to help the team overcome the obstacles causing a delay.

Finally, the program manager will follow the program life cycle and ensure that once the program has achieved success in all of the identified benefits, it follows a clear and concise closeout procedure including releasing resources, developing lessons learned, documenting program risks, and determining financials for the effort. This closeout process provides input for both the program manager and the organization in achieving repeatable success for additional efforts undertaken.

Quite often the program manager drives the productivity of the various project teams by focusing on milestones, approving phase gates (go/no-go decision points), and communicating successes out through the stakeholder community. The program manager can assist project managers in meeting their established objectives by continuously communicating the objectives, rewarding high achievers, and acknowledging the successes of the project teams themselves. Program managers operate at a higher level than project managers and when necessary can balance resources among projects to ensure that each effort achieves its defined goals and objectives.

DRIVING PROJECT MANAGEMENT THROUGH BETTER PROGRAM MANAGEMENT

Program management is a formalized management approach that leverages process and knowledge areas to create a framework managing multiple projects and operational efforts that together are capable of achieving benefits that individual projects could not achieve on their own. As such, program management is based on the project management processes as a foundation and interacts with the project manager at predefined points in the project life cycle. While projects are managed leveraging the *PMBOK* containing five process groups and ten knowledge areas, program management builds on these using the same five process groups, but leverages five domains for program management.

As such, program managers hold a certain level of expectation of their project managers. They expect proper *PMBOK* processes to be implemented and reported on in such a way so as to enable the program to report on its composite pieces. In addition, the program relies on projects to identify not only risks to their effort but also enterprise- and program-level risks that could impact other projects or the program as a whole. These risks are escalated to the program manager to be part of the program risk management matrix and managed accordingly.

A project not following PMI standards would not conform to the program governance set up by the program manager and would eliminate the ability of the program manager to effectively manage the program and ensure stakeholders that the benefits would be realized in a timely and efficient manner. In addition, risks may not be identified, documented, or escalated to the program manager, leaving the program manager with very little insight into any issues being encountered by the project and therefore unable to assist with additional resources, SMEs, or reallocation of dollars. Program risks not communicated up to the management level are often repeated among multiple projects and can be eliminated only when a trend is identified. If project managers consistently report both risks and strategies, the program management team can identify root causes and potentially implement strategies to avoid the duplication of a risk in subsequent project efforts.

In reality, program and project management are so tightly integrated that program management cannot be effective if project management standards are not followed.

DISTINCTIONS BETWEEN PROGRAM AND PROJECT MANAGEMENT

Although program and project management are dependent upon one another, there are also definitive differences in both approaches. Project management is focused on delivering a single unique product or service and upon completion transitioning that product or service to an operational team for user support. Program management is the management of multiple projects and operational efforts in an attempt to achieve results not available from managing a single project. Thus program management includes operational support and will span the life of multiple projects, at times managing numerous projects in various points of their program life cycle. The program manager is also responsible for communicating with the stakeholder community as a whole. Whereas a project may have a subset of stakeholders, and the project manager will communicate with those stakeholders as milestones are achieved, the program manager needs to be able to speak to the stakeholder community as a whole, identifying the project milestones, benefits that have been realized, and potential program-level risks.

More specifically:

Scope: Projects have defined objectives and in many cases the scope is progressively elaborated gaining greater information as the effort advances.

Programs have a larger scope and provide deliverables in the form of benefits delivered through projects and operational efforts.

Change: Project Managers expect change and implement processes to manage changes and ensure that they do not affect scope, cost, schedule, or quality.

Programs expect change from both internal as well as external sources and must manage scope to ensure that benefits are realized in a timely and cost effective manner. When a change is presented a program manager and change control board can evaluate many alternatives. If a change is outside the scope of one project it could be shifted to another project that will be initiated at a later date, depending on the deliverables inherent risks of the change, and project schedules established.

Planning: Project managers leverage progressive elaboration, analogous, bottom-up, or top-down analysis to determine the approach to be taken in meeting the project objectives.

Programs develop initial high level plans and project placeholders to be filled in by Project Managers as they go through their planning process. The Program Management plan will cover a much longer term and program schedules will include multiple projects, operational efforts, and transition points than an individual project and may include Level of Effort (LOE) tasks for supporting operational endeavors.

Management: Project Managers manage the project team on a temporary basis with the intent of meeting the project objectives.

Program Managers manage the Program staff, Project Managers, and indirectly the project team members by providing vision, leadership, and escalation paths for staff.

Success: Project Managers measure success through client satisfaction, cost, schedule, and quality of product or service delivered.

Program Managers measure success by the degree to which the benefits realized meet the needs of stakeholders and are aligned with the objectives of the program.

Monitoring: Projects monitor and control the work of producing the product or service through earned value analysis and customer satisfaction.

Programs are monitored and controlled through monitoring each component project through earned value analysis, and deliverables measured against the benefits defined for the Program. (PMI 2013b)

GOVERNANCE AS SUPPORT FOR PROGRAM MANAGEMENT AND NOT AS A BATON

Program governance is the formalized management approach that program managers establish to ensure that each project contained within the program leverages equivalent processes and procedures in the management and reporting on project progress. While the governance aspect of program management can been seen as oversight or a negative "big brother" management style, in reality it provides program managers with insight into each project and the progress being made.

As I mentioned earlier, a project manager has a single, tangible product or service that he or she focuses on delivering. The program manager is responsible for that deliverable as well as any number of additional projects depending on the size of the program. With the insight gained through program governance, the program manager can often offer assistance to individual projects by balancing resources across multiple projects, reallocating funds, or even potentially taking some pieces of the deliverables off the plate of the project manager to assist in ensuring project success. In addition, a successful program manager leverages governance to provide insight into the progress of a project and can then offer his or her support to the project manager if he or she sees issues being encountered.

Finally, whereas a project manager is concerned with the risks that affect his or her project specifically, a program manager may be able to recognize a pattern from the identified risks and can work with multiple project managers to develop strategies to resolve the risk at the root cause rather than each time it is realized in a project. At the end of the day, the program manager is not an adversary of the project manager; rather, they both need each other to achieve success, and as I have harped on just a bit, communication is the key to success.

Information communicated to the program manager on project status, risks, schedules, and costs is not data that is passed directly on to management or clients; instead it must be both interpreted and evaluated by the program manager as it relates to the overall program scope. From this information, the program manager must identify strategies that he or she can leverage to assist the projects in managing risks and meeting the project objectives, schedules, and costs. All of this information is collected through the governance of the program. Without that detailed information, a program manager will remain uninformed and unable to assist a project manager until potentially too late to be beneficial.

CASE STUDY

You have just been hired as a program manager to lead an international effort implementing production facilities around the world. Each facility will be located in a different country and requires individual architecture, government regulations, local technology capabilities, and infrastructure support. The organization has defined its strategic objectives of (1) increasing production output around the world by 30 percent, (2) increasing sales through globalization by 25 percent, and (3) increasing customer satisfaction through consistent and reliable quality standards.

Your program will be instrumental in achieving these objectives by increasing production capacity and facilitating sales efforts. In addition, the production facilities will implement a set of quality standards defined by the organization and will leverage a Lean Six Sigma approach eliminating waste and increasing efficiency. Once the production facilities are built, staffed, and fully functional, the program will transition them to functional management reporting to the chief operations officer.

Although you will be based in the United States, you will have a flexible travel budget and be able to work with project teams around the world as needed. In addition, travel expenses for team members are included in your budget. Telephone and video conference technologies are in place and available to you whenever needed.

Much of the program will include local subcontractors hired for work efforts such as building roads and buildings, furnishing facilities, wiring offices, setting up production equipment, and supplying raw materials. All facilities must comply with local regulations as well as U.S. standards.

As a program manager, you will be responsible for defining objectives for each project, implementing financial and program controls, reporting to the board of directors and executive management, resolving issues internally within the organization, planning the overall implementation, developing strategies for handing off completed facilities for operational control, and managing program risks. You will be providing monthly reports to executive management and reporting on a quarterly basis to the board of directors.

There are no corporate processes in place for project management or program management, and the organization has been unsuccessful in earlier attempts to expand globally. While you have budget dollars for hiring additional staff within the program, there is no corporate support for programs and very limited documented processes for procurement and financial management. The program will operate with a $50 million budget and estimates place the cost of all the facilities at $45 million. Because your teams are geographically diverse, many project managers will speak multiple languages including English, but English will be their second language. Team members may or may not speak English but will speak in the local dialect.

CHAPTER QUIZ

1. Which domain of program management is the most useful in achieving HPTs?
2. How do you use program governance without slowing down project efforts and wasting project managers' time on unnecessary or superficial tasks?
3. How are program risks defined?
4. What portion of the stakeholder community is a program manager responsible for? Which stakeholders are project managers responsible for?
5. How do you resolve conflicts between project managers and ensure that open communication exists?
6. What are the most important considerations for building HPTs?
7. What leadership style do you use the most? When would you use an alternate style? How would you know when to transition from an alternate leadership style to your choice of styles?
8. What impact can leadership styles have on teams?
9. How does a program manager ensure success on his or her next program effort?
10. If no process is in place, what materials would you use to jump-start a process effort?

DISCUSSION QUESTIONS

1. With no program or project management process in place, how would you work with project managers to achieve success?
2. Project teams are located throughout the world. How would you approach team building?
3. What goals could you set to achieve HPTs for each of the projects?
4. What strategy for communication would you use with project managers?
5. Considering all teams are in different time zones, when would you schedule regular meetings?
6. How would you ensure that project managers were sharing their project achievements and challenges with one another?

7. Assuming you would use some form of reward system for HPTs exceeding expectations, how would you approach rewarding teams or individuals?
8. With multiple sites and projects being started all over the world, how would you manage your time to be most effective?
9. What internal staffing positions would help you achieve greater success?
10. Would you implement any education or training programs, and how would you handle the logistics of project manager locations?
11. What ways could technology be leveraged to increase your chances of program success?

18

Conclusion

> Men make history and not the other way around. In periods where there is no leadership, society stands still. Progress occurs when courageous, skillful leaders seize the opportunity to change things for the better.
>
> **—Harry Truman**

The focus of this book has not been on the program management guidelines as outlined in the *Standard for Program Management*, third edition (PMI 2013b). There are a number of books being published on that subject, and the latest version from PMI is one of the better books that you could possibly find to learn the process. Instead, this book has been on leadership and more specifically how program managers lead teams to become HPTs that are capable of achieving together much more than they could as individual players.

While projects are temporary endeavors undertaken to create a unique product, service, or result (PMI 2013a, 2), programs are the creation of a new result, product, or service through the culmination of multiple projects, operational efforts, and process change that alone could not achieve the same benefits as when combined into a single effort (PMI 2013b). Programs will contain multiple efforts whereas a project is focused more on a single result. Project managers will drive the creation of a single result, service, or product based on the objectives defined for the effort. Although they will be aware of overall program benefits, their job is drive toward project success and will focus on those deliverables. On the other hand, a program manager is responsible for multiple projects and operational efforts and will be able to tie the needs of each project together into a single plan for achieving program benefits. Project managers by definition are myopic, focusing on the work that needs to be done for the project, whereas program managers operate at a much higher level, looking across multiple efforts. While program management understands the project management standards and requires them to be applied, the program manager follows a different approach, which contains five domains: strategy alignment, program benefits management, program stakeholder management, program governance, and program life cycle management. These five domains ensure that the program manager actively aligns the program benefits with the strategic objectives defined to achieve the benefits of the program, manages stakeholder expectations, and ensures that underlying components are following and reporting on consistent processes, and that the program follows a logical approach for life cycle management.

To be successful as a leader, a program manager must have the ability to drive team performance, communicate effectively with both stakeholders and team members, ensure that all efforts are managed in a consistent approach, and work with sponsors to validate that programs remain in alignment with the organizational strategic benefits. Furthermore, a program manager must have the ability to draft and communicate the benefits that will be and are achieved throughout the program life cycle. Longer-term programs will leave stakeholders feeling frustrated with the short-term pain of working with outdated technology or infrastructure, but benefits management and stakeholder management are used to ensure that all program stakeholders are consistently made aware of the long-term benefits and are able to celebrate short-term accomplishments as they are achieved.

While the mechanics of program management are crucial, the art of leadership is vital to achieving success for programs. Leadership is the ability to motivate team members to, for a period of time, set aside their own personal motivators, agendas, and drivers, and instead work toward a common goal that is established by the program manager. It is through leadership that teams come together and strive for success even when it is difficult to come by. Leaders do not force, drive, or dominate teams; rather, they communicate, empower, and challenge teams in such a way that the team members willingly contribute their time and effort to the goals of the program. Program managers begin with establishing a clear and consistent vision that can be used with both teams and stakeholders that identifies the benefits to the program and why each component is necessary to achieve the overall success. Program visions are clarified and repeated often enough that every team member will be able to repeat the vision, embrace and support it, and understand how its achievement will improve the organization and help meet its long-term goals.

While a program will have multiple component projects and potentially operational efforts to achieve the benefits established, the teams can range from program-level team members to those participating in project efforts. Regardless of the contribution or the part of the program that members are working on, they are all part of the whole program team and all must be motivated to communicate, innovate, create, and develop a level of confidence that the work they produce will help the program (and therefore the organization) achieve strategic goals.

Teams can be individuals thrown together for the purpose of a project, initiative, or program, but when managed properly they can form HPTs that exceed expectations, overcome challenging obstacles, leverage innovation, and approach risk from an aggressive stance. HPTs rely on one another to accomplish goals and form communication channels that enable them to perform at a dramatically increased rate over their counterpart organizations. HPTs generally leverage their personal competitive spirit and become emotionally involved in achieving success, whereas teams not driven often simply go through the motions, anticipating program/project failure.

Each team, regardless of how they were brought together, will follow a set of team development phases. However, not all teams will be successful in moving through all stages of team development and can get hung up on one or the other. The five stages of team development as defined by Tuchman (1965) are:

1. Forming
2. Storming
3. Norming
4. Performing
5. Adjourning (or mourning) (384)

When a team hits the performing stage, they are entering the realm that HPTs operate in. A team that is performing but is driven further by invoking personal emotion, competitive spirit, and a clear, focused, but challenging goal can transition to an HPT with proper leadership.

At the same time, teams may never move out of the storming or norming stage if the leadership does not facilitate open communication, encourage positive conflict resolution, and empower teams to invest their personal emotions in the achievement of success. While these teams may achieve some level of success, the effort will not be a positive one and may result in team members resenting the project and holding negative feelings against either team members or the leadership. They generally do not feel that they have achieved success and are often demoralized by the effort.

An HPT will achieve the objectives, have a sense of satisfaction, and reach the adjourning phase more as a mourning phase, sad that the effort is over and the team will be breaking up to work on other projects/programs. This final phase of team development is one of the determinants of how effective the leadership was. When a team reaches adjourning and they cannot wait to get away from each other, leadership failed in some way and may not have resolved conflicts, opened communications, or done an effective job of assigning roles and responsibilities.

Program management is more of an art than a science. Following the guidelines from PMI will make you a program manager; but employing leadership, driving toward a common vision, opening communication channels, and managing conflict will all help with building teams to be more effective. However, achieving HPTs requires a concerted effort that empowers the staff and motivates them to invest their personal ambition in the achievement of the objectives. A program manager can achieve success for the effort, but one who builds HPTs will exceed expectations, building teams that outperform and accomplish goals that will be unlikely for others. An HPT will enjoy the program and will want to work with the leader and team members on future efforts. HPTs enjoy the work that is performed and take great pride in the outcome of the program.

Therefore program management guidelines and mechanics when complemented by leadership can achieve a level of satisfaction, achievement, and success that is not possible by those just going through the motions. These kinds of leaders will drive organization improvement by leading through example and will create a safe and secure "bubble" for their teams to operate within.

Unfortunately, there is not a single leadership model that will be useful in every situation. While most of us think that we are transformational, empowering everyone around us and making everyone feel good about their job, the reality is that we all employ different styles of leadership. Whether it be command-and-control (authoritative), transactional, dark, or transformational leadership, all leaders have a management style that is most comfortable to them. Of course based on the situation at hand, employing different leadership styles can be beneficial to the success of the program. The key to leveraging situational leadership is recognizing the need, consciously choosing the approach, knowing when the approach has met the goals, then moving on to a different style of leadership that is more effective in the long term.

I personally feel comfortable in a transformational leadership style, but often find that the situation at hand requires a more authoritative or transactional approach. It can be disturbing to find that your leadership style is not complementary to team development, but through this recognition we can learn to employ other strategies that make us more effective.

Hopefully, this book has helped you understand the value of leadership and how it can generate greater performance, morale, and team development, driving performance that others are simply not capable of. Leadership is one of the most vital components in any form of management, whether leading projects, programs, or organizations. Understanding the leadership approaches and leveraging them consciously and intentionally drives people to work together in a more effective manner.

References

Alon, I., Higgins, J., 2005. Global Leadership through emotional and cultural intelligences, Kelley School of Business, Indiana University.

Barrett, R. 1998. *Liberating the Corporate Soul: Building a Visionary Organization*. Woburn, MA: Butterworth-Heinemann.

Bass, B. M. 1981. *Stogdill's Handbook of Leadership*. New York: Macmillan.

Bass, B. M. 1990. "From Transactional to Transformational Leadership: Learning to Share the Vision." *Organizational Dynamics* 18 (3): 19–32.

______. 1999. "Two Decades of Research and Development in Transformational Leadership." *European Journal of Work and Organizational Psychology* 8 (1): 9–32.

Blanchard, K., and M. O'Connor. 1997. *Managing by Values*. San Francisco: Berrett-Koehler.

Burns, J. M. 1978. *Leadership*. New York: Harper & Row.

Cashin, J., Crewe, P., Desai, J., Desrosiers, L., Prince, J., Shallow, G. & Slaney, S. 2000. Transformational Leadership. Retrieved August 3, 2006 from http://www.mun.ca/educ/ed4361/virtual_academy/campus_a/aleaders.html

Chemers, M. M. 2000. "Leadership Research and Theory: A Functional Integration." *Group Dynamics* 4 (1): 1089–2699.

Ciulla, J. B. 2003. *The Ethics of Leadership*. Belmont, CA: Thomson Wadsworth.

Conger, J. A. 1990. 'The dark side of leadership', *Organizational Dynamics*, 19, 44–45.

Conger, J. A., and R. N. Kanungo. 1988. "The Empowerment Process: Integrating Theory and Practice." *Academy of Management Review* 13: 471–83.

Covey, S. R. 1991. *Principle-Centered Leadership*. New York: Summit.

Day, D., S. J. Zaccaro, and S. M. Halpin, eds. 2004. *Leader Development for Transforming Organizations: Growing Leaders for Tomorrow*. Mahwah, NJ: Erlbaum.

Despain, J., and J. Converse. 2003. *And Dignity for All: Unlocking Greatness with Values-Based Leadership*. Upper Saddle River, NJ: Financial Times Prentice Hall.

Drucker, P. F. 2001. *The Essential Drucker: Selections from the Management Works of Peter F. Drucker*. New York: Ehrhart & Klein.

Elton, M. 1949. *Hawthorne and the Western Electric Company: The Social Problems of an Industrial Civilization*. New York: Routledge.

Enderle, G. 1987. *Sicherung des Existenzminimums im nationalen und internationalen Kontext: Eine wirtschaftsethische Studien*. Bern, Switzerland: Haupt.

Engelbrecht, A. van Aswegen, and C. Theron. "The Effect of Ethical Values on Transformational Leadership and Ethical Climate in Organizations," South African Journal of Business Management, vol. 36, no. 2, 2005, pp. 19–26. - See more at: http://www.chausa.org/publications/health-progress/article/may-june-2007/transformational-leadership#sthash.J3Vp04dv.dpuf

Erickson, A., J. Shaw, and Z. Agabe. 2007. "An Empirical Investigation of the Antecedents, Behaviors, and Outcomes of Bad Leadership." *Journal of Leadership Studies* 1 (3): 26–43.

Fairholm, G. W. 1998. *Perspectives on Leadership: From the Science of Management to Its Spiritual Heart*. Westport, CT: Quorum Books.

Foucault, M. 1970. *The Order of Things: An Archaeology of the Human Sciences*. New York: Vintage Books.

Freud, S. 1923. *The ego and the id*. SE, 19: 1-66.
Gabaro, J., and J. Kotter. 2005. Managing your boss, *Harvard Business Review*, January 2005 Vol 1.
Gabriel, Y., and A. Carr. 2002. "Organizations, management and psychoanalysis: An overview." *Journal of Managerial Psychology* 17 Iss: 5, pp. 348–365.
Gallagher, E. G. 2002. "Leadership: A Paradigm Shift." *Management in Education* 16 (3): 24–29.
Gardner, H. 1990. *Art Education and Human Development*. Occasional Paper 3, September. Los Angeles: Getty Center for Education in the Arts.
Goff, J. L., and P. Goff. 1991. *Organizational Co-dependence: Causes and Cures*. Boulder: University Press of Colorado.
Hewett, E. 1889. *Elements of Psychology*. Cincinnati, OH: Eclectic Press.
Hogan, R., G. Curphy, and J. Hogan. 1994. "What We Know about Leadership: Effectiveness and Personality." *American Psychologist* 00: 493–504.
Holmberg, I., and J. Ridderstrale. 2000. "Sensational Leadership." In *Management 21C*, edited by S. Chowdhury, 000–000. London: Financial Times Prentice Hall.
Howell, J. 1992. "The Ethics of Charismatic Leadership: Submission or Liberation." *Academy of Management Executive* 6 (2): 43–54.
Howell, J. M., and C. A. Higgins. 1990. "Champions of Technological Innovation." *Administrative Science Quarterly* 35 (2): 317–41.
Ilan, A., and James M. Higgins. 2005. "Global Leadership Success through Emotional and Cultural Intelligences." *Business Horizons* 48 (6): 501–12.
Jennings, E. 1961. "The Anatomy of Leadership." *Management of Personnel Quarterly* 1 (1): 2–9.
Johnson, N., and T. Klee. 2007. "Passive-Aggressive Behavior and Leadership Styles in Organizations." *Journal of Leadership and Organizational Studies* 14 (2): 130–42.
Judge, T. A., J. A. LePine, and B. L. Rich. 2006. "Loving Yourself Abundantly: Relationship of the Narcissistic Personality to Self and Other Perceptions of Workplace Deviance, Leadership, and Task and Contextual Performance." *Journal of Applied Psychology* 91 (4): 762–76.
Katzenbach, Jon R., and Douglas K. Smith. 2003. *The Wisdom of Teams: Creating the High-Performance Organization*. New York: HarperCollins.
Khoo, H. S., and G. S. J. Burch. 2008. "The 'Dark Side' of Leadership Personality and Transformational Leadership: An Exploratory Study." *Personality and Individual Differences* 44: 86–97.
Kirkpatrick, S. A., and E. A. Locke. 1991. "Leadership: Do Traits Matter?" *Academy of Management Executive* 5 (2): 48–60.
Knight, C., and D. Dyer. 2005. "Ten Traits of Effective Leaders." *Harvard Management Update* 10 (10): 3–6.
Kotter, J. 1990. "What Leaders Really Do." *Harvard Business Review* 68 (3): 103–11.
Kouzes, J. M., and B. Z. Posner. 1995. *The Leadership Challenge*. San Francisco: Jossey-Bass.
______. 2007. *The Leadership Challenge*. San Francisco: Wiley.
Lasch, C. 1980, The Culture of Narcissism: American life in an age of diminishing expectations, Warner Books, New York.
Leithwood, K. & D. Jantzi. 2000. The effects of transformational leadership on organizational conditions and student engagement with school. *Journal of Educational Administration*, 38(2), p. 112.
Levinson, H. 1972. *Organizational Diagnosis*. Cambridge, MA: Harvard University Press.
______. 1976. *Psychological Man*. Boston: Levinson Institute.

______. 1981. *Executive*. Cambridge, MA: Harvard University Press.
Lewin, K. 1936. *Principles of Topological Psychology*. New York: McGraw-Hill.
Lewin, K., and R. Lippitt. 1938. "An Experimental Approach to the Study of Autocracy and Democracy: A Preliminary Note." *Sociometry* 1: 292–300.
Lewin, K., R. Lippitt, and R. K. White. 1939. "Patterns of Aggressive Behavior in Experimentally Created Social Climates." *Journal of Social Psychology* 10: 271–301.
Luthans, F., and B. Avolio. 2003. "Authentic Leadership: A Positive Development Approach." In *Positive Organizational Scholarship*, edited by K. S. Cameron, J. E. Dutton, and R. E. Quinn, 241–58. San Francisco: Berrett-Koehler.
Maio, G. R., J. M. Olson, L. Allen, and M. Bernard. 2001. "Addressing Discrepancies between Values and Behavior: The Motivating Effect of Reasons." *Journal of Experimental Social Psychology* 37: 104–17.
Maslow, A. 1943. "A Theory of Human Motivation." *Psychological Review* 50: 370–96.
McIntosh, G., and S. Rima. 1997. *Overcoming the Dark Side of Leadership: The Paradox of Personal Dysfunction*. Grand Rapids, MI: Baker Books.
Meglino, B., E. Ravlin, and C. Adkins. 1992. "The Measurement of Work Value Congruence: A Field Study Comparison." *Journal of Management* 18 (1): 33–43.
Meindl, J. R., and S. B. Ehrlich. 1987. "The Romance of Leadership and the Evaluation of Organizational Performance." *Academy of Management Journal* 30 (1): 91–109.
Mitroff, I. 2005. "Crisis Leadership: Seven Strategies of Strength." *Leadership Excellence* 22: 11.
Morf, C. C., and F. Rhodewalt. 2001. "Unraveling the Paradoxes of Narcissism: A Dynamic Self-Regulatory Processing Model." *Psychological Inquiry* 12: 177–96.
Morris, M. H., and D. F. Kuratko. 2002. *Corporate Entrepreneurship*. Mason, OH: South-Western College.
Mumford, M. D., S. J. Zaccaro, F. D. Harding, T. Owen Jacobs, and E. A. Fleishman. 2000. "Leadership Skills for a Changing World: Solving Complex Social Problems." *Leadership Quarterly* 2 (1): 11–35.
O'Toole, J. 1996. *Leading Change: The Argument for Values-Based Leadership*. New York: Ballantine Books.
Pienaar, C. 2008. "The Role of Self-Deception in Leadership Effectiveness." *South African Journal of Psychology* 39 (1): 133–41.
Prilleltensky, I. 2000. "Bridging Agency, Theory and Action: Critical Links in Critical Psychology." In *Critical Psychology: Voices for Change*, edited by T. Sloan, 67–81. London: Macmillan.
Project Management Institute. 2013a. *A Guide to the Project Management Body of Knowledge (PMBoK Guide)*. 5th ed. Newton Square, PA: Project Management Institute.
______. 2013b. *The Standard for Program Management*, 3rd ed. Newton Square, PA: Project Management Institute.
Robson, C. 2002. *Real World Research*, 2nd ed. Oxford: Blackwell.
Rokeach, M. 1973. *The Nature of Human Values*. New York: The Free Press.
Schaef, A. W., and D. Fassel. 1998. *The Addictive Organization*. San Francisco: Harper & Row.
Schein, E. 2004. *Organizational Culture and Leadership*. 3rd ed. San Francisco: Jossey-Bass.
Stogdill, R. 1974. *Handbook of Leadership: A Survey of Theory and Research*. New York: The Free Press.
Taylor, F. 1911. Principles of Scientific Management, Harper and Row Publishers, Inc. New York, NY.
Thornbury, N. 2006. *Entrepreneurial Strategies for Innovation and Growth*.
Topping, P. 2002. *Managerial Leadership*. New York: McGraw-Hill.

Treviño, L. K., M. Brown, and L. Pincus-Hartman. 2003. "A Qualitative Investigation of Perceived Executive Ethical Leadership: Perceptions from Inside and Outside the Executive Suite." *Human Relations* 56 (1): 5–37.

Waelder, R. 1967. "Inhibitions, Symptoms and Anxiety: Forty Years Later." *The Psycho-Analytic Quarterly* 36 (1): 1–36.

Weber, M. 1947. *The Theory of Social and Economic Organization*. New York: The Free Press.

Zaleznik, A. 1977. "Managers and Leaders: Are They Different? *Harvard Business Review* 55: 47–60.

______. 1989a. *The Managerial Mystique*. New York: Harper & Row.

______. 1989b. "The Mythological Structure of Organizations and Its Impact." *Human Resource Management* 28 (2): 267–78.

Index

D

E

O

P

U

V

W

Printed in the United States
by Baker & Taylor Publisher Services